Fourier Optics in Image Processing

SERIES IN OPTICS AND OPTOELECTRONICS

Series Editors: Robert G W Brown, University of California, Irvine, USA

E Roy Pike, Kings College, London, UK

This international series includes books on all aspects of theoretical and applied optics and optoelectronics. Titles in this series range in level from advanced textbooks for graduate students to research monographs on specialized subjects including: Nonlinear Optics and Laser Radiation, Interferometry, Waves, Crystals and Optical Materials. These titles will also meet the requirements of scientists and engineers working in optics, optoelectronics and related fields in the industry.

Handbook of Solid-State Lighting and LEDs
Zhe Chuan Feng (Ed.)

Nanophotonics and Plasmonics: An Integrated View
Dr. Ching Eng (Jason) Png and Dr. Yuriy Akimov

Handbook of Optoelectronic Device Modeling and Simulation: Lasers, Modulators, Photodetectors, Solar Cells, and Numerical Methods – Volume Two
Joachim Piprek (Ed.)

Handbook of Optoelectronic Device Modeling and Simulation:
Fundamentals, Materials, Nanostructures, LEDs, and Amplifiers – Volume One
Joachim Piprek (Ed.)

Handbook of GaN Semiconductor Materials and Devices
Wengang (Wayne) Bi, Hao-chung (Henry) Kuo, Pei-Cheng Ku, and Bo Shen (Eds.)

Handbook of Optoelectronics, Second Edition: Applied Optical Electronics – Volume Three
John P. Dakin and Robert Brown (Eds.)

Handbook of Optoelectronics, Second Edition: Enabling Technologies – Volume Two
John P. Dakin and Robert Brown (Eds.)

Handbook of Optoelectronics, Second Edition: Concepts, Devices, and Techniques – Volume One
John P. Dakin and Robert Brown (Eds.)

Optical Microring Resonators: Theory, Techniques, and Applications Vien Van
Thin-Film Optical Filters, Fifth Edition
H. Angus Macleod

Optical MEMS, Nanophotonics, and Their Applications
Guangya Zhou and Chengkuo Lee (Eds.)

For more information about this series, please visit: https://www.crcpress.com/Series-in-Optics-and-Optoelectronics/book-series/TFOPTICSOPT

Fourier Optics in Image Processing

By
Neil Collings

CRC Press
Taylor & Francis Group
Boca Raton London New York

CRC Press is an imprint of the
Taylor & Francis Group, an **informa** business

CRC Press
Taylor & Francis Group
6000 Broken Sound Parkway NW, Suite 300
Boca Raton, FL 33487-2742

First issued in paperback 2020

© 2019 by Taylor & Francis Group, LLC
CRC Press is an imprint of Taylor & Francis Group, an Informa business

No claim to original U.S. Government works

ISBN-13: 978-0-367-57153-5 (pbk)
ISBN-13: 978-1-4987-6068-3 (hbk)

Version Date: 20180514

Visit the Taylor & Francis Web site at
http://www.taylorandfrancis.com

and the CRC Press Web site at
http://www.crcpress.com

To my wife, Lydia, and my children. Ines, Ernest, and Emma.

Contents

Preface

Fourier optics as a discipline grew between 1940 and 1960. Duffieux completed a monograph on the subject in 1944, which was privately published after the war in 1946. The work gained little attention outside France, until Born and Wolf called attention to it in their text, *Principles of Optics* (1959). The lack of a rigorous mathematical basis to the book has led to criticism from some quarters. In particular, the lack of the theory of generalized functions, such as the delta function, was seen as a grave omission. However, the originality of the viewpoint expressed leads to an agreement from all scholars that this book marked the beginning of Fourier optics in the spatial domain. The subject was opened up to a wider community when *Introduction to Fourier Optics*, by Joseph W. Goodman, was published in 1968. In subsequent editions of this book, an introduction to some of the application areas of this discipline was provided. In particular, two types of optical correlators, the Vander Lugt Correlator and the Joint Transform Correlator, which were conceived in the 1960s, are described in approximately 20 pages of the third edition of this book. A second edition of Duffieux's book was published in 1970 by Masson (Paris) and an English translation appeared in 1983. Fourier optics is now the cornerstone for the analysis of diffraction, coherence, and holography, as well as specialized topics such as wavefront control, and propagation through random media.

The optical correlator will be presented here in more detail than in Goodman's book. It is a sensor which detects whether a scene contains a specified content. In certain situations it can replace a human operator: for example, where there is a possibility of fatigue reducing the judgement, or where there is the possibility of overload (large number of visual tasks required). In other cases, the operator cannot be present and quick decisions should be made. For a number of these situations, it is relevant how compact and low power this sensor can be made, so that we shall take a historical view of compact correlators and look at what systems engineering can currently do in this respect. The apogee of optical correlators was probably the New Product Award to the

Institut National d'Optique in Canada during the 1999 Conference on Lasers and Electro-optics (CLEO). My work in this field began in 1983, and resulted in a correlator system which is described in the Appendix of my book [62]. Since that time, I have never been away from Fourier Optics in one form or another.

Many people have helped directly or indirectly with this book. My daughter, Dr. Ines E. Collings, guided me through my initial stumbles with LaTeX scripting. Professor Rafie Mavaddat performed the experiment resulting in Fig. 5.2. The boundless enthousiasm and aptitude of my previous and present employers, Professor Bill Crossland and Dr. Jamie Christmas, and the rigour of my earlier employer, Professor Dr. Rene Dandliker, have been of inestimable importance. Finally, I am thankful to the small but enthusiastic following of optical processing, many of whom I have met at conferences around the globe.

It is hoped that the book will provide a solid grounding in an applied field of Fourier Optics, so that students and researchers can find stimulation for practical work. Fourier optics is no longer a "hot" topic which eased my task because I did not have to keep a running watch on current literature. However, I have tried to present up-to-date device and components options which are accessible for the experimenter. Much work remains to be done, with the aim of developing an integrated system which is competitive. Perhaps the two main topics which require more research in order to make optical correlator systems a commercial prospect, are optical packaging and good applications for the technology. Despite a number of attempts to make an integrated optical system assembly, nothing has been found which can compete with multichip modules, high-density multiway connectors, or multilayer printed circuit boards (PCBs). I have looked at some of the attempts at a packaging solution and applications of this discipline in the final chapter.

Acronyms

For the reader's convenience, we introduce here the acronyms that we will use in different places in the remainder of the book. They are listed in order of appearance.

HVS	Human Visual System
CPU	Central Processing Unit
NN	Neural Network
HOG	Histogram of Oriented Gradients
SIFT	Scale-invariant Feature Transform
CNN	Convolutional Neural Network
SVM	Support Vector Machine
GPU	Graphics Processing Unit
FT	Fourier Transform
DFT	Discrete Fourier Transform
FFT	Fast Fourier Transform
FPGA	Field Programmable Gate Array
I/O	Input/Output
ASIC	Application Specific Integrated Circuit
DSP	Digital Signal Processor
DCT	Discrete Cosine Transform
EASLM	Electrically Addressed Spatial Light Modulator
SBWP	Space Bandwidth Product
MTF	Modulation Transfer Function
CMOS	Complementary Metal Oxide Semiconductor
NMOS	N-type Metal Oxide Semiconductor
TFT	Thin Film Transistor
SRAM	Static Random Access Memory
CCD	Charge-Coupled Device
DRAM	Dynamic Random Access Memory
DMD	Digital Micromirror Device
DLP	Digital Light Processing
LCOS	Liquid Crystal on Silicon

rms	root mean square
fps	frames per second
FSC	Frame Sequential Colour
SXGA	Super Extended Graphics Array
ITO	Indium Tin Oxide
FF	Fill Factor
AR	Aperture Ratio
MOSFET	Metal-Oxide-Semiconductor Field Effect Transistor
CMP	Chemical-Mechanical Polishing
NLC	Nematic Liquid Crystal
PAN	Planar Aligned Nematic
ECB	Electrically Controlled Birefringence
VAN	Vertically Aligned Nematic
TH	Tilted Homeotropic
SLM	Spatial Light Modulator
HFE	Hybrid Field Effect
TN	Twisted Nematic
MTN	Mixed Twisted Nematic
OCB	Optically Compensated Bend
FLC	Ferroelectric Liquid Crystal
FLCOS	Ferroelectric Liquid Crystal on Silicon
FLCD	Ferroelectric Liquid Crystal Device
MEMS	Micro-Electro-Mechanical System
MOEMS	Micro-Opto-Electro-Mechanical System
MMA	Micromirror Array
DM	Deformable Mirror
GLV	Grating Light Valve
NB	Normally Black
NW	Normally White
OASLM	Optically Addressed Spatial Light Modulator
LCLV	Liquid Crystal Light Valve
BSO	Bismuth Silicon Oxide
DOE	Diffractive Optical Element
CGH	Computer Generated Hologram
IFT	Inverse Fourier Transform
MSE	Mean Squared Error
IFTA	Iterative Fourier Transform Algorithm
SA	Simulated Annealing
PWS	Plane Wave Spectrum
RIE	Reactive Ion Etch

OTF	Optical Transfer Function
MTF	Modulation Transfer Function
USAF	U.S. Air Force
PSF	Point Spread Function
SR	Strehl Ratio
ADC	Analogue-to-Digital Conversion
PC	Personal Computer
DR	Dynamic Range
EMVA	European Machine Vision Association
SLVS-EC	Scalable Low Voltage Signalling Embedded Clock
CMOSIS	CMOS Image Sensor
sCMOS	scientific CMOS
SNR	Signal to Noise Ratio
AR	Anti-reflection
LVDS	Low-Voltage Differential Signalling
PSD	Position Sensitive Detector
PS	Profile Sensor
DL	Diode Laser
LED	Light Emitting Diode
FWHM	Full Width Half Maximum
VCSEL	Vertical Cavity Surface Emitting Laser
TIR	Total Internal Reflection
SLED	Super Luminescent LED
VLC	Vander Lugt Correlator
JTC	Joint Transform Correlator
ffp	front focal plane
bfp	back focal plane
HRM	Holographic Recording Material
DE	Diffraction Efficiency
PR	Photorefractive
HSF	Holographic Spatial Filter
MF	Matched Filter
JPS	Joint Power Spectrum
BS	Beamsplitter
TFLOPS	Teraflops (floating point operations per second)
HM	Holographic Memory
cps	correlations per second
XOR	Exclusive OR
MOC	Miniaturized Optical Correlator
MROC	Miniature Ruggedized Optical Correlator

GOC	Greyscale Optical Correlator
FOM	Figure of Merit
DC	Discrimination Capability
POC	Phase-only Correlation
SVD	Singular Value Decomposition
cdf	cumulative distribution function
pdf	probability density function
GOC	Greyscale Optical Correlator
ROC	Receiver Operating Characteristic
DET	Detection Error Trade-off
DR	Detection Rate
FAR	False Alarm Rate
PCE	Peak-to-Correlation Energy
PSR	Peak-to-Sidelobe Ratio
MF	Matched Filter
POF	Phase-only Filter
BPOF	Binary Phase-only Filter
GLs	Grey Levels
MED	Minimum Euclidean Distance
SPR	Statistical Pattern Recognition
SDF	Synthetic Discriminant Function
PCA	Principal Component Analysis
ICA	Independent Component Analysis
ECP	Equal Correlation Peak
MOF	Mutual Orthogonal Filter
MACE	Minimum Average Correlation Energy
MACH	Maximum Average Correlation Height
PIFSO	Planar Integration of Free-Space Optics
SAR	Synthetic Aperture Radar
ATR	Automatic Target Reognition
DNA	Deoxyribonucleic Acid
bp	base pair
ROI	Region of Interest
DOG	Difference of Gaussians
MV	Machine Vision
CID	Charge Injection Device
RGB	Red, Green and Blue
UAV	Unmanned Autonomous Vehicle
DEM	Digital Elevation Map

Introduction

CONTENTS

1.1 COMPUTER VISION

Computer vision is a project to reproduce the remarkable performance of the human visual system (HVS) in a machine. It is both important and difficult. Some generic and particular applications are: medical diagnosis; quality inspection; robotics; driverless cars; missile guidance; car number plate recognition; pedestrian detection; and tracking people in a crowd. Image recognition and identification in complex scenes are the areas most actively researched. Computer vision is conventionally divided into three functional areas: low-level, mid-level, and high-level vision. This division follows a hierarchical approach to the HVS. Low-level vision covers the segmentation of the image taking place between the eye and the cortex. Mid-level vision, involving colour, form, and movement, occurs in the prestriate cortex. Finally, high-level vision, such as recognition, occurs in the inferior temporal cortex. The neural signals do not travel in a single direction from the low- to the high-level: there is a feedback of neural activity between the three divisions. Experimental data show that visual comprehension of images is fast [206]. This has been explained on the basis of modelling using feature extraction operating within a hierarchical feedforward neural network architecture [234].

In the early days of computer vision, an equivalence was established between picture functions which are analytically well-behaved and digital picture functions which are discrete arrays of numbers which can be processed on digital computers [221]. The digitisation of picture functions takes place in cameras, which will be discussed in Chapter 4. The camera samples the picture due to its inherent pixellisation. When the sampling is fine-grain, then a high-resolution digital picture function results. Prior to this, optical systems for processing pictures were considered because of the limited parallelism of the digital computers of the day. In conventional (von Neumann) digital computers a central processing unit (CPU) is connected to memory and to the input/output devices by a system bus. The computing power of the von Neumann computer derives from the CPU which can perform mathematical operations on high precision data at a high speed. This is inefficient for operations on digital picture functions which benefit from architectures with greater parallelism [15]. The connectionist, or neural network (NN), approach to computer vision involves highly interconnected systems of simple thresholding elements. In the NN approach, the power of the structure is the connection network between the thresholding elements rather than the processing elements themselves. An analogue weight is associated with each connection and the value of the weight is refined during the learning phase, where the NN is presented with exemplars (i.e., a known input/output pairing). The development of algorithms for the weight update (learning) is an important part of research in this field. In the early days of NNs, the ambitious goal was to develop a system which could generalise. For example, after a limited learning phase on the classification of a range of images, the NN could be presented with an image which had not been previously presented and it would classify it correctly. Limited success at generalisation was an initial setback in the connectionist field. An early attempt at a single layer NN was the perceptron [220]. It received a poor reception because its computational capability was limited. However, this single layer unit was an important first step in combining analogue weights; thresholding neurons; and a simple learning algorithm. It has led to further developments, such as the multilayer perceptron, and the synthetic discriminant functions which will be presented in Section 7.7.2.

In order to improve on the limited capabilities of a single layer network, multilayer networks were also researched, together with backpropagation learning. However, the real impetus to advances in the field of computer vision came with the availability of high throughput computers and good, large databases. Computing power nowadays is deployed to process a large number of features over the whole of the image. Examples of this are the feature detectors which record particular groupings of grey levels of pixels using neighbourhood operations, such as the Haar wavelet features in Viola, Jones [269]; histogram of oriented gradients (HOG) [64]; and scale-invariant feature transform (SIFT) [164]. These are analogous to the low-level processing of the HVS. Learning proceeds with the help of a high-level support vector machine (SVM) classifier [265]. Alternatively, it is claimed that convolutional NNs (CNNs) can extract good features which allow generalization using a general-purpose learning procedure [152]. Up to 40 million images per day can be learned with high-end graphics processing units (GPUs), where memory is tightly coupled to processing elements in an architecture with high parallelism.

1.2   TEMPLATE MATCHING AND FEATURE EXTRACTION

Template matching is the comparison of a number of stored patterns (templates) with the image until the template is located within the image and its positional coordinates are retrieved. This is a successful technique for retrieving the pattern in a noisy image when there is an exact match between the stored pattern and the pattern in the image. However, due to the variability of a given image (for example, the pose, illumination, and expression of a face), template matching using the whole face as a template is less successful. Smaller size templates can be used more effectively. For example, the eyes on the face can be represented by two black circles separated by a defined distance. This is a template for the eyes that can be rotated by varying angles and searched for in the image. This will allow detection of the frontal face at varying angular pose.

There are a number of strategies which can be employed to perform template matching. Here we illustrate the cross-correlation approach, which consists of sliding the template across the image and computing the product of the pixels of the template and the image at each position of the template. For a PxQ template, $T(P, Q)$, and an LxM image,

$I(L, M)$, the product is given by

$$C(l, m) = \sum_{p=0}^{P-1} \sum_{q=0}^{Q-1} T(p, q) I(l + p, m + q) \qquad (1.1)$$

where $C(l, m)$ is one pixel of the cross-correlation matrix, $C(L, M)$. Therefore, PQ multiplications which have to be summed are computed for each pixel of the cross-correlation matrix. This is approximately $2PQ$ arithmetic operations, which are floating point operations (flops), if the pixels are represented by floating point numbers. Therefore, the total number of flops required to find the PxQ template in the LxM image is approximately $2PQ(L-P+1)(M-Q+1)$, when the template is slid across all possible locations in the image. In this calculation, the template never protrudes beyond the perimeter of the image.

Whereas template matching is expensive at both the computational level and in terms of memory resources, the alternative approach of feature extraction is relatively inexpensive on both counts. Feature extraction tackles the problem by defining pixel neighbourhood features of the pattern rather than the pattern itself, and searching for these features. Some prominant types of feature detectors were mentioned in Section 1.1. The high computational cost of searching for a template in the image is replaced by the smaller computational and memory costs of simple neighbourhood operations on pixels. However, the number of these simple operations is increased dramatically compared with the template matching approach. Therefore, the small unit computational cost is amplified by the large number of units involved. This would extend the processing time beyond real-time operation. However, algorithms have been developed which allowed the detection of frontal faces within a 384 by 288 window at a rate of 15 frames per second using a conventional computer with a 700 MHz CPU [269].

1.3 FOURIER OPTICS

Fourier optics takes its name from Jean-Baptiste Fourier (1768-1830) who, in 1807, published a memoir on the conduction of heat in a circular ring. He formulated a partial differential equation to describe the conduction which he solved as a series of trigonometrical functions. This is now known as the Fourier series representation of a periodic function. The decomposition of non-periodic functions is accomplished with the Fourier transform (FT). The application of this transform in optics arises due to the phenomenon of optical diffraction, which is an aspect of the wave nature of light.

1.3.1 Diffraction

An intuitive mathematical treatment of optical diffraction is that based on the Huygens-Fresnel principle [90]. According to this principle, light propagates as a wavefront and, in order to calculate successive positions of the wavefront, it is sufficient to propagate wavelets from secondary sources on the wavefront. Wavelets are spherical waves which are solutions of the Helmholtz equation which governs the complex amplitude of monochromatic optical disturbances propagating through free space. The resulting wavefront in the output plane is the envelope of the wavelets from these secondary sources:

$$U(\mathbf{r}') = \frac{1}{i\lambda} \int_S U(\mathbf{r}) \frac{e^{ikR}}{R} \cos\left[\theta(\mathbf{r})\right] ds, \qquad (1.2)$$

where $U(\mathbf{r})$ is the complex amplitude of the wavefront at position vector $\mathbf{r}$, $R = |\mathbf{r}' - \mathbf{r}|$ and $\theta(\mathbf{r}' - \mathbf{r})$ is the angle between the normal to the plane of the aperture and $(\mathbf{r}' - \mathbf{r})$ (Figure 1.1); λ is the wavelength and $k = 2\pi/\lambda$ is the spatial angular frequency or magnitude of the wave vector. The integration is performed over the area of the aperture. It can be viewed as the summation of elementary Huygens wavelets multiplied by the amplitude of the electric field at the origin of the wavelet in the plane of the aperture.

The following approximations to the propagation distance, R, in the exponential factor give the formulae for Fresnel and Fraunhofer diffraction:

$$\begin{aligned} R &= \left[z_0^2 + (x' - x)^2 + (y' - y)^2\right]^{\frac{1}{2}} \\ &\simeq z_0 + \frac{(x' - x)^2 + (y' - y)^2}{2z_0} \qquad (1.3) \\ &\simeq z_0 + \frac{(x'^2 + y'^2)}{2z_0} + \frac{(x^2 + y^2)}{2z_0} - \frac{x'x + y'y}{z_0} \qquad (1.4) \\ &\simeq z_0 - \frac{x'x + y'y}{z_0} \qquad (1.5) \end{aligned}$$

Equation (1.4) is the approximation used for Fresnel diffraction where

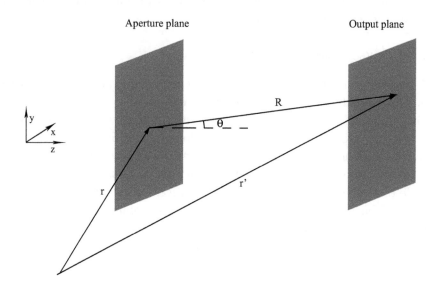

Figure 1.1 Coordinate system for variables in the Rayleigh-Sommerfeld diffraction integral (Equation (1.2))

the fourth power terms in the coordinate variables are ignored. Equation (1.5) is the approximation used for Fraunhofer diffraction where the quadratic terms of the coordinate variables are ignored. Under the Fresnel approximation, Equation (1.4),

$$U(x',y',z_0) = \frac{1}{i\lambda z_0} exp\left[ik\left(z_0 + \frac{(x'^2+y'^2)}{2z_0}\right)\right].$$
$$\int\int U(x,y,0) exp\left[ik\left(\frac{(x^2+y^2)}{2z_0} - \frac{x'x+y'y}{z_0}\right)\right] dxdy$$

(1.6)

Under the Fraunhofer (or paraxial) approximation, Equation (1.5),

$$U(x',y',z_0) = \frac{e^{ikz_0}}{i\lambda z_0} \int\int U(x,y,0) exp\left[ik\left(-\frac{x'x+y'y}{z_0}\right)\right] dxdy$$

(1.7)

and the FT of the aperture multiplied by a constant appears in the output plane. The intensity distribution in the output plane is the complex modulus squared of the FT. The quadratic terms can beignored when the path length to the output plane, z_0, is large compared to the variables, x, x', y, and y'. In order to achieve this result in an optical system, a lens placed between the two planes reduces the path length requirement. When it is placed so that it is intermediate between the two planes and their separation is 2f, then both the quadratic phase factors are cancelled by the quadratic phase function of the lens [90]. This is known as 'the Fourier transforming property of a lens'. The focal length of the lens cannot be reduced below the level at which aberrations reduce the quality of the FT. The design of FT lenses for acceptably low levels of aberration is considered by several authors [270], [125], and [284].

Fourier Transform

The formula for the propagation from the aperture plane to the output plane, Equation (1.6), is the FT of the aperture function multiplied by a constant. The two-dimensional FT is a transform between the spatial domain, x, y, and the frequency domain, f_x, f_y. The frequency plane appears at the output plane or the focal plane of the lens, where the spatial coordinates are related to f_x, f_y by $x' = f_x \lambda z_0$, $y' = f_y \lambda z_0$, and $x' = f_x \lambda f$, $y' = f_y \lambda f$, respectively. In the case of the lens, the complex amplitude of the electric field can be expressed as a function of two spatial coordinates, and the transform, $\hat{U}(f_x, f_y)$, is a function of the two frequency coordinates

$$\hat{U}(f_x, f_y) = \iint_{-\infty}^{\infty} U(x,y) exp[-2\pi i(f_x x + f_y y)] \, dx dy \qquad (1.8)$$

When propagating from the output plane back to the aperture plane, the inverse transform is used. This is defined as follows

$$U(x,y) = \iint_{-\infty}^{\infty} \hat{U}(f_x, f_y) exp[2\pi i(f_x x + f_y y)] \, df_x df_y \qquad (1.9)$$

The definition of a frequency, $f_x = x'/\lambda f$, avoids the need for a normalizing constant in front of the integral sign in these equations. If the input function, $V(x,y)$, is real and symmetric about the origin, so that $V(x,y) = V(-x,y) = V(x,-y) = V(-x,-y)$, then $\hat{V}(f_x, f_y)$ is a real function which is also symmetric about the origin, so that $\hat{V}(f_x, f_y) = \hat{V}(-f_x, f_y) = \hat{V}(f_x, -f_y) = \hat{V}(-f_x, -f_y)$, as follows

$$
\begin{aligned}
\hat{V}(f_x, f_y) &= \iint_{-\infty}^{\infty} V(x,y) exp[-2\pi i(f_x x + f_y y)] \, dx dy \\
&= \iint_{-0}^{\infty} V(x,y)(exp[-2\pi i(f_x x + f_y y)] + exp[2\pi i(f_x x + f_y y)]) \, dx dy \\
&= 2 \iint_{-0}^{\infty} V(x,y) cos[-2\pi(f_x x + f_y y)] \, dx dy \qquad (1.10)
\end{aligned}
$$

1.3.2 Spatial frequency of a grating

If the aperture plane is a periodic structure, such as a transmission grating, then the aperture function, $U(x, y)$, in Equation 1.8, can be written as the product of a periodic function, $g(x, y)$, multiplied by a function expressing the extent of the aperture. If $g(x, y)$ is expressed as a series of sine and cosine terms, then the Fourier integral will give the coefficients of these terms. The series representation of the periodic function is known as the Fourier series representation of the grating. The diffraction pattern of this grating is a sequence of spots in the output plane. The central spot is called the zero order, and the first order spots on either side of the central spot represent the fundamental spatial frequency of the grating, and the higher order spots represent the harmonics. The intensity of each spot is the complex modulus squared of the coefficient of the corresponding harmonic of the Fourier series. The particular case of a Ronchi grating, which is a square wave grating with unity mark/space ratio, is discussed in Section 4.2. The spatial frequency of a grating is the reciprocal of the period of the grating, with equivalent units of cycles per mm or linepairs per mm. Therefore, the finer the pitch of the grating, the more widely spaced will be the spots in the output plane. For a grating of pitch, Λ, the spatial frequency is $\eta = 1/\Lambda$.

1.3.3 Convolution and correlation

The convolution of two functions, f and h, is defined as

$$g(x') = \int_{-\infty}^{\infty} f(x)h(x' - x)\mathrm{d}x = f \otimes h \qquad (1.11)$$

where $\otimes$ is a shorthand symbol for the convolution operation. The Fresnel approximation, Equation (1.6), can be viewed as a 2D convolution between the aperture function and a Gaussian function

$$U(x', y', z_0) = \frac{e^{ikz_0}}{i\lambda z_0} \iint U(x, y, 0)\, exp\left[ik\left(\frac{(x - x')^2 + (y - y')^2}{2z_0}\right)\right].$$
$$\mathrm{d}x\mathrm{d}y \qquad (1.12)$$

Sometimes it is possible to simplify a convolution integral if the FTs of both functions are known. The convolution theorem states that the

Fourier transform of the convolution of two functions is equal to the product of their individual Fourier transforms,

$$\int_{-\infty}^{\infty} g(x')e^{-iux'}\,\mathrm{d}x' = \int_{-\infty}^{\infty} f(x)e^{-iux}\,\mathrm{d}x \int_{-\infty}^{\infty} h(x'-x)e^{-iu(x'-x)}\,\mathrm{d}x'$$

(1.13)

The correlation of two functions, f and h, is defined as

$$c(x') = \int_{-\infty}^{\infty} f(x)h^*(x-x')\mathrm{d}x = f \odot h^*$$

(1.14)

where $\odot$ is a shorthand symbol for the correlation operation. The equivalent version of Equation (1.13) is

$$\int_{-\infty}^{\infty} c(x')e^{-iux'}\,\mathrm{d}x' = \int_{-\infty}^{\infty} f(x)e^{-iux}\,\mathrm{d}x \int_{-\infty}^{\infty} h^*(x-x')e^{-iu(x-x')}\,\mathrm{d}x'$$

(1.15)

An important property of the spatial domain correlation is that of shift invariance. If the input function is displaced along the x-axis by D, then the correlation is displaced by the same distance, which permits target tracking in an optical correlator

$$\int_{-\infty}^{\infty} f(x+D)h^*(x-x')\mathrm{d}x = \int_{-\infty}^{\infty} f(x'')h^*(x''-D-x')\mathrm{d}x'' = c(x'+D)$$

(1.16)

Symbol conventions for convolution and correlation do vary in the literature. For example, the opposite convention is found in [90]. The present convention is in keeping with, for example, the $\otimes$ used for convolution in [244]. Older literature uses the $*$ symbol for convolution and does not distinguish between convolution and correlation. The correlation of one function with itself is called the autocorrelation, and with another function the cross-correlation. The principal distinction between convolution and correlation is that the function $h(x)$ is inverted about $x = 0$ in the former case, but not in the latter case. When $h(x)$ is symmetric and real, convolution and correlation are identical.

An important application of the convolution theorem is to recover $f(x)$ from $g'(x)$ in situations where $h(x'-x)$ and its FT are known. This is commonly required in imaging systems when the effect of the transfer function of the system must be removed (Section 4.2). This is known as deconvolution.

1.3.4 Fourier shift theorem

The Fourier transform given by Equation (1.7) appears centred on the optical axis. In order to translate the transform laterally by small amounts perpendicular to the axis, it is necessary to add linear phase terms to the phase of the complex amplitude in the aperture plane. The complex amplitude is expressed as a phasor

$$U(x, y, 0) = a(x, y, 0)e^{ig(x,y,0)}. \tag{1.17}$$

If a linear phase ramp in x is added, then

$$U(x, y, 0) = a(x, y, 0)e^{ig(x,y,0)+irx}, \tag{1.18}$$

where r is the phase ramp in rad/m.

Inserting the modifed complex amplitude into Equation (1.7) gives

$$U(x', y', z_0) = \frac{e^{ikz_0}}{i\lambda z_0} \int \int U(x, y, 0) \, exp \left[ik \left(-\frac{x'x + y'y}{z_0} \right) + irx \right].$$
$$dxdy \tag{1.19}$$

With a change of variable of $x'' = x' - \frac{rz_0}{k}$, then

$$U(x'', y', z_0) = \frac{e^{ikz_0}}{i\lambda z_0} \int \int U(x, y, 0) \, exp \left[ik \left(-\frac{x''x + y'y}{z_0} \right) \right].$$
$$dxdy \tag{1.20}$$

The centre of the Fourier transform is displaced to a new position and the extent of the displacement is given by $-\frac{rz_0}{k}$. Conversely, translations in the aperture plane give rise to corresponding phase changes in the output plane. This property of the Fourier transform is known as the Fourier shift theorem.

1.4 DIGITAL TECHNIQUES

The preceding section described the analogue Fourier transform which results from the process of optical diffraction. The discrete Fourier transform (DFT) is used for numerical calculation of the Fourier transform. It is defined by

$$F(u, v) = \sum_{l=0}^{L-1} \sum_{m=0}^{M-1} f(l, m) \, exp \left[-2\pi i \left(\frac{lu}{L} + \frac{mv}{M} \right) \right] \tag{1.21}$$

where $f(l,m), F(u,v)$ are matrices of complex values of dimension (L, M). The summations are implemented in one dimension initially followed by the second dimension. In addition, there is an implied periodicity in the data series, for which it is assumed that f(L)=f(0). For example, a 1D vector, $f(l) = [0, 1, 0, -1]$, lists the four components $f(0), f(1), f(2)$, and $f(3)$;and it is assumed that $f(4) = 0$. It has a Fourier transform, $F(u) = 2i[0, -1, 0, 1]$. The first component, $F(0)$, is the coefficient of zero spatial frequency, and is called the DC or zero order component; it is the sum of the components of $f(l)$. The second component, $F(1)$, is the coefficient of the fundamental spatial frequency, one period within $0, L$. The third and fourth components, $F(2)$ and $F(3)$, are the coefficients of the second and third harmonics of the fundamental spatial frequency, two and three periods within $0, L$. Since $f(l)$ is antisymmetric about the midpoint $L/2$, the fundamental and harmonics are sin waves, multiplied by $-i$ from the expansion of the exponential. Moreover, since $f(l)$ is a real function, these imaginary components are anti-symmetric.

1.5 TRADE-OFF WITH DIGITAL (NUMERICAL) APPROACHES

Early work on computer vision was concerned with the definition of picture functions and digital picture functions [221]. The former are "indistinguishable from" analytically well-behaved functions and can be Fourier transformed. The latter are "piecewise constant" picture functions which can be manipulated as matrices inside a digital computer. At that time, picture functions were captured on photographic film for coherent optical processing, or on active devices such as photoconductive/electroluminescent sandwiches for incoherent processing. A summary of a full range of differences between analogue and digital processing are listed in Table 1.1. The digital computer allows full flexibility of programming and is capable of high precision arithmetic calculations. However, optics offers the possibility of processing images from the real world or from solid state databases with no interruption. In addition, the analogue optical processor can be made smaller and with lower power consumption. Analogue processing has seen a recent resurgence in the research arena due to the launch of a Defense Advanced Research Projects Agency (DARPA) program in the Unconventional Processing of Signals for Intelligent Data Exploitation (UPSIDE) in 2013.

The computational complexity of template matching was addressed in Section 1.2. In order to reduce the amount of computation, schemes such as multi-resolution matching (or hierarchical template matching) can be employed, where the sampling of the image and template are progressively reduced in order to reduce complexity. This is appropriate for templates with low spatial frequency content which is retained in the version with reduced sampling. For templates where the high spatial frequencies are key to the matching process, the complexity reduction can be achieved by using Fourier techniques. The cross-correlation can be calculated as a product of Fourier transforms (Equation 1.15) followed by an inverse Fourier transform (Equation 1.9).

Once the numerical computation has been reduced to a succession of Fourier transforms, the calculation can be accelerated by the use of the fast Fourier transform (FFT). In the FFT, the elements of the matrix which is to be Fourier transformed are paired up and computed in parallel for a set of values of the exponential factor. The latter is called the twiddle factor, which is computed and stored in memory. The computation of a pair of elements with one of the set of twiddle factors results in a pair of products for the succeeding parallel computation. This is called a butterfly operation due to the shape of the computational flow. The paired product implementation is known as a radix-2 FFT implementation. If this is done in successive stages of the calculation, then the complexity of the computation is reduced from fourth order in the matrix dimension to quadratic order. In addition to the number of floating point operations required, there is a need for tightly coupled memory (low access time). Increased parallelism can be employed at the cost of increased power consumption. The digital hardware which is currently used to perform 2D FFTs includes FPGA (field programmable gate array) and GPU. The speed and performance of these two approaches are impressive, although the power dissipation is relatively large. The typical power dissipation of an FPGA implementation is 1W static; 2.5W dynamic; 1W input/output (I/O); and 0.8W transceiver, giving a total of 5.3W. The power consumed on the board of the NVidia GPU TK1 is 3.7W (idle) and 10.2W(active).

A large research effort is devoted to reducing power consumption for digital image processing. Traditionally, an application specific integrated circuit (ASIC) was designed with significant reduction in power dissipation. A number of comparison websites exist for the FFT chips which have been fabricated, both commercially and in academia. The speed of a 1024-point FFT together with power dissipation and number

TABLE 1.1   Comparison of digital and analogue optical image processing

Digital	Analogue optical
Fully programmable	Restricted programmability
Serial memory	Parallel memory
High precision	Limited precision
Digital I/O	Possibility of analogue I/O

of chips required are the standard comparison criteria. Digital signal processor (DSP) chips have been developed for the FFT of, principally, audio signals. They can also be used for the discrete cosine transform (DCT) of images. The DCT can approximate lines in the image with fewer coefficients than the DFT. It is used in image compression where an algorithm based on the DCT has been defined by the Joint Photographic Experts Group (JPEG) committee of the International Standards Organization. For the CNN, a recent example of an ASIC is Eyeriss [57], with a power consumption of 278 mW.

Because of the heavy computational load required, traditional CPUs and DSPs can be rapidly overburdened. The preferred option for digital processing is the FPGA with a specialist FFT core. However, for large matrix sizes, corresponding to high-resolution images, specialist algorithms have to be developed in order to make efficient use of the off-chip memory resources [3]. The trade-off with analogue processing is in terms of precision, power consumption, cost, and size. For the small sensor product, the analogue approach holds many advantages. This is precisely the area where optical Fourier transforms could have a large impact, in the sensors used for data collection and processing.

GLOSSARY

Feature extraction: Searching for pixel neighbourhood features within the scene.

Fourier optics: The field of optics where the spatial Fourier transforming aspect of optical diffraction is exploited.

Fourier series: The representation of a periodic function by the sum of a set of simple oscillating functions based on the fundamental spatial frequency and its harmonics.

Fourier transform: The representation of a continuous function by a spectrum of frequencies in the spatial frequency domain.

Neural network: A highly interconnected system of simple thresholding elements.

Resolution: The granularity of a digital picture function.

Spatial frequency: The natural unit of the spatial Fourier series and transform.

Template matching: Searching for a stored image within a scene.

CHAPTER 2

Spatial light modulators

CONTENTS

2.1 OPTICS OF DIGITAL PICTURE FUNCTIONS

Any discrete representation of a signal/image on an electrically addressed spatial light modulator (EASLM) is a sampled version of the real-life signal/image. This has two important corollaries: that the sampling creates a periodicity in the Fourier representation of the signal/image; and that the resolution is limited by the number of pixels in the EASLM.

2.1.1 Replications and apodisation

Even if there is no periodicity in the signal, the periodicity of the sampling creates a periodicity in the Fourier transform. The periodic replications of the Fourier transform are known as higher orders. The mathematical basis of higher orders can be understood using the principles developed in Chapter 1. When the aperture plane is defined by a physical electrooptic device such as an EASLM, the input image is sampled at the pixel repeat of the device and each sample is multiplied by the physical extension of the pixel. The total extent of the input, $U(x, y, 0)$, is then bounded by the aperture of the EASLM. This limitation is known as apodisation. For a unit width rectangular aperture on the x-axis, the notation $rect(x)$ is used, with corresponding spectral function, $sinc(f_x)$, where $sinc(f_x) = sin(\pi f_x)/(\pi f_x)$ (Figure 2.1).

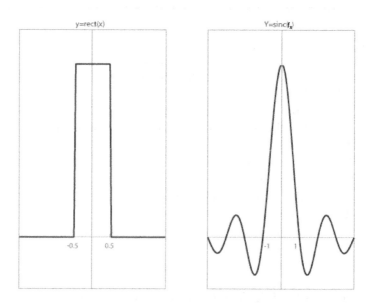

Figure 2.1 Graphical representation of a unit width 1D aperture and its Fourier transform

The *rect* function is important for the representation of both the pixel aperture and the device aperture. In each case, it is normalised by the aperture values, a, b and $L\Delta x, M\Delta y$, respectively, where L, M are the number of pixels in the x, y axes (Figure 2.2). For example, an aperture, a, is represented by $rect(x/a)$, which drops to zero at $-0.5a$ and $0.5a$. The corresponding spectral function is $sinc(f_x a)$, which drops to zero at $-1/a$ and $1/a$. Therefore, the first zeroes of the spectral function appear at spatial coordinates $-\lambda z_0/a$ and $\lambda z_0/a$ in a plane at distance, z_0, from the EASLM.

The delta function, $\delta(x)$, is important for locating, mathematically, the spatial position of each pixel. It is an example of a generalized function [157] which is defined by its effect on a second function, $g(x)$,

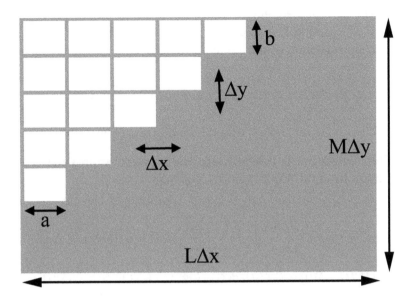

Figure 2.2 Labelling of the pixel sizes, pixel repeats, and total dimensions of the EASLM

as follows

$$\int_{-\infty}^{\infty} \delta(x - x_0) g(x) \, \mathrm{d}x = g(x_0)$$

In particular, when $g(x)$ is the complex exponential function, the FT is generated

$$\int_{-\infty}^{\infty} \delta(x - x_0) e^{-i f_x x} \, \mathrm{d}x = e^{-i f_x x_0}$$

A *comb* function can be used to represent the infinite sum of δ functions separated by a repeat distance of x_0 along the x axis [90]. In particular,

$$comb(x) = \sum_{l=-\infty}^{\infty} \delta(x - l x_0).$$

The FT of the spatial *comb* function is a spatial frequency *comb* function

$$\int_{-\infty}^{\infty} \sum_{l=-\infty}^{\infty} \delta(x - l x_0) e^{-i f_x x} \, \mathrm{d}x = \frac{1}{x_0} \sum_{l=-\infty}^{\infty} \delta(f_x - l/x_0) = \frac{1}{x_0} comb(f_x),$$

$$(2.1)$$

where the spatial frequency δ functions are separated by a repeat distance of $1/x_0$ along the f_x axis.

For the sampling of the image, the delta function notation is used,

$$U(x,y,0) = rect(x/L\Delta x)rect(y/M\Delta y) \sum_{l=0}^{L-1}\sum_{m=0}^{M-1} U(x,y,0).$$

$$rect(x/a)\delta(x - l\Delta x)rect(y/b)\delta(y - m\Delta y) \quad (2.2)$$

This expression is inserted into the paraxial approximation of the diffraction integral, Equation (1.7), giving

$$U(x',y',z_0) = \frac{e^{ikz_0}}{i\lambda z_0} \int\int \sum_{l=0}^{L-1}\sum_{m=0}^{M-1} U(x,y,0)rect(x/L\Delta x).$$

$$rect(y/M\Delta y)rect(x/a)rect(y/b)\delta(x - l\Delta x)\delta(y - m\Delta y)). \quad (2.3)$$

$$exp\left[ik\left(-\frac{x'x + y'y}{z_0}\right)\right] dxdy$$

Due to the windowing effect of the full aperture *rect* functions, $rect(x/L\Delta x)$ and $rect(y/M\Delta y)$, the limited summation over the δ functions in Equation (2.3) can be replaced by *comb* functions. The Fourier transform of the spatial *comb* function is a spatial frequency *comb* function, Equation (2.1).

The sampling converts the analogue Fourier transform into a discrete Fourier transform (Section 1.4). The input image, $U(x,y,0)$, is a LxM matrix of values, with individual elements $U(l\Delta x, m\Delta y, 0)$. Its Fourier transform is given by

$$\hat{U}(f_x,f_y) = \sum_{l=0}^{L-1}\sum_{m=0}^{M-1} U(l\Delta x, m\Delta y)exp\left[-2\pi i\left(f_x l\Delta x + f_y m\Delta y\right)\right]$$

$$(2.4)$$

Therefore,

$$U(x',y',z_0) = \frac{e^{ikz_0}}{i\lambda z_0}[\hat{U}(f_x,f_y) \otimes comb(f_x)comb(f_y))].$$

$$sinc(f_x a)sinc(f_y b) \quad (2.5)$$

In this expression, the spatial domain is discretised in space domain coordinates $(l\Delta x, m\Delta y)$, and the frequency domain is similarly discretised in frequency domain coordinates $(u/L\Delta x, v/M\Delta y)$. The latter expression has been replaced by (f_x, f_y) for convenience. The analogue Fourier transform created, Equation (2.5), is the discrete Fourier transform, $\hat{U}(f_x, f_y)$, replicated by the *comb* functions and apodised by the *sinc* functions. Apodisation here refers to the gradual attenuation of $\hat{U}(f_x, f_y)$ as the value of f_x or f_y is increased. When the zero of the *sinc* function coincides with the higher order generated by the *comb* function, then the replicas of the analogue Fourier transform are centred on these zeroes. This occurs when the fill factor of the pixel (see Section 2.3) is 100%, which has not yet been achieved in a practical device.

2.1.2 Space bandwidth product

The total number of pixels is known as the space bandwidth product (SBWP) of the EASLM. Alternative acronyms used in the literature, such as SBP and SW, refer to the 1D SBWP which is the number of pixels in one dimension of the EASLM. Their use is also extended to lenses and systems, so the SW of the system [163] or the SBWP of a lens [97] is also found. The bandwidth used in the SBWP is the reciprocal of the pixel repeat in mm. The SBWPs of some EASLMs are given in Table 2.1.

The microdisplays listed in Table 2.1 have been designed with a particular commercial application in mind. One aspect of this design is the array size. For square pixel microdisplays, the EASLM SBWP follows the same format as the relevant TV aspect ratio. Those microdisplays with widths 1024, 1280, and 2048 pixels follow the 4:3 aspect ratio of standard TV. Those with a width of 1920 pixels follow the 16:9 aspect ratio of high definition TV. Those with a width in excess of 4000 pixels are intended for application in ultra-high-definition TV. Some devices which have been developed exclusively for spatial light modulation are the Meadowlark and Spatial Light Machines devices.

In order to improve the performance of a coherent optical processing system, it is important to increase the SBWP of the EASLM. This is usually limited by cost as far as the EASLM is concerned. However, a further limitation can be introduced by the lens used to form the Fourier transform. As the SBWP of the EASLM is increased, the aperture of the lens must be increased and the demands on the quality of the lens increase also. If the pixel repeats, $\Delta x, \Delta y$, are decreased as the SBWP is increased, then the aperture required will remain constant; however, the extent of the spectral function formed in the focal plane of the lens will increase. The quality of the lens is also critical for maintaining an accurate representation of the spectral function in the Fourier plane. Aberrations, in particular field curvature, will increase the spot size in the Fourier plane, if they are not corrected. Correct lens design, following the references cited in Section 1.3.1, is important. In order to assess the quality of a given lens, the modulation transfer function (MTF) is used. The MTF will be discussed in Chapter 4.

2.2 EASLM TECHNOLOGY AND PERFORMANCE PARAMETERS

EASLMs are two-dimensional pixellated array devices which act as transducers between analogue or digital electrical signals and Fourier optical systems which employ coherent light beams. A coherent light beam can be modulated in amplitude, phase, and polarization and all these modes can be used in image processing systems. In amplitude modulation, the matrix, $U(l\Delta x, m\Delta y)$, is an array of real numbers. In phase modulation, the matrix is an array of unity amplitude phasors. Finally, in polarization modulation, the matrix can be decomposed into two matrices of phasors, which represent the complex amplitudes of the electric field along two orthogonal axes in the (x, y) plane. The EASLMs which will be discussed in this chapter cover two device technologies, liquid crystal and deformable mirror. The liquid crystal devices are both transmissive and reflective, whereas the deformable mirrors are reflective. Semiconductor electroabsorption modulators are also fabricated in 128x128 pixel format for high speed optical communication systems [273].

2.2.1 EASLM technology

All the devices discussed in this chapter are electrically addressed using an active backplane. The backplane is the addressing structure of row and column electrodes, and 'active' refers to the presence of one or more transistors at each intersection of row and column. The reflective devices are based on integrated circuits in single crystal silicon. These are fabricated using either complementary metal oxide semiconductor (CMOS) or N-type metal oxide semiconductor (NMOS) processing. The CMOS processing is usually categorized by the minimum feature size, e.g., 0.18 μm. In general, the smaller the feature size, the smaller the pixel size of the EASLM, because the circuitry can be accommodated within a smaller area. The full wafer of silicon (up to 300 mm diameter) provides a large number of backplanes. Fuller details of the backplane technology are given in [10]. The transmissive devices are liquid crystal devices driven by either a CMOS backplane which has been thinned sufficiently to become transparent, or thin film transistors (TFTs). The thinned backplane is attached to a suitable transparent substrate, such as sapphire, in order to provide mechanical support, whereas the TFT device is built up by processing layers deposited onto a transparent substrate, such as glass. The latter process has been perfected for large size displays, such as computer monitors and televisions.

Three types of address structure have been used to charge the pixels of the active CMOS backplane: static random access memory (SRAM); charge-coupled device (CCD); and dynamic random access memory (DRAM). SRAM is a digital addressing architecture which was originally developed for the Texas Instruments (TI) Digital Micromirror Device (DMD) [104]. The CCD address was developed for a CCD-addressed liquid crystal light valve [72]. The CCD devices were fabricated using the older NMOS fabrication technology. DRAM addressing originated in 1971 as a method of driving TFT displays [130]. The two addressing techniques now used are the SRAM for the digital light processing (DLP) device and a majority of liquid crystal on silicon (LCOS) devices; and the DRAM for the remainder of the LCOS devices. They are both fabricated using CMOS processing. The SRAM address is often referred to as digital address, and the DRAM address is known as analogue address. Digital address consists of a repetition of 'bit planes' where the optical modulation of each pixel is either 'on' or 'off'. This provides excellent voltage uniformity across the active area of the panel, and consequently good modulation uniformity across the active area of the EASLM. In the DLP, the micromirrors have a low inertia and respond to the digital address with little delay. When the same digital driving is applied to a liquid crystal layer where inertial effects are greater, the response is slower. The modulation of the electrooptic effect is the convolution of the digital drive waveform with the response time of the liquid crystal. The match between the switching time of the nematic liquid crystal and the repetition rate of 'bit planes' is critical for good performance with the SRAM backplane. Modulation of the reflected or transmitted light is observed when the liquid crystal responds to the voltage of the individual 'bit planes' rather than the time averaged root mean square (rms) voltage. This effect can be ameliorated either by driving the backplane faster or using a slower liquid crystal, for example a thicker layer or low temperature operation. However, a slower liquid crystal response time, either by increasing the layer thickness or reducing the temperature, is not desirable because the speed of the device is reduced. Higher clock frequencies entail increased power dissipation in the driver.

The analogue address backplane was developed for TFT displays because the speed of the electronics did not allow multiple scans to provide the grey scale, as in the digital address scheme. It employs a capacitor at each pixel which is charged once per frame. The DRAM address was created so that, in one pass of the array, each pixel element was furnished with an analogue charge which was stored in the pixel capacitor. This provides a stable value for the pixel voltage, limited only by voltage droop due to charge leakage. In order to minimize the direct current (dc) level on the liquid crystal (called dc balancing), alternating current (ac) drive based on frame inversion is employed. This reduces the available drive voltage to at most one half of the available voltage, unless the front electrode of the device is switched between the upper and lower drive voltages. This is known as alternating front electrode drive. The liquid crystal response time can now be short so that the frame rate of the device can be optimised. The DRAM advantage for Fourier optical systems is that each pixel in the array can have the correct grey level concurrently for a given period of time, so that the correct interference of the incident light wave takes place. The drawbacks are drift- and channel-dependent variations of drive voltages which need to be compensated. The slew rate of the analogue signal can cause ringing, nonlinear distortion, noise, and echoes. The DRAM pixel also requires a large capacitor to store the voltage level of the drive waveform. The switching speed between neighbouring grey levels is also longer than the SRAM architecture where the same drive voltages are used for all grey levels.

The main differences between microdisplays and EASLMs are in terms of wavefront distortion, physical size, modulation capability, and cost. Coherent optical systems benefit from low wavefront distortion devices in order to reduce aberrations. Small aperture devices also favour low wavefront distortion, whereas the etendue requirements of displays favour larger aperture devices. Consequently, the pixel sizes of EASLMs are typically smaller than those of microdisplays. Microdisplays are designed for good contrast amplitude modulation at a speed consistent with video display rate. A greater variety of modulation modes are acceptable for Fourier optics application, with phase and complex modulation especially of interest. Finally, a coherent optical system can usually bear a higher cost EASLM than an equivalent microdisplay.

The speed at which an EASLM operates is a critical parameter which impacts directly the speed of the image processing system. In the case of a microdisplay device, the frame speed is usually 60 frames per second (fps). Each frame is commonly composed of a number of sub-frames. The sub-frames are used, for example, to sequence the three colours in a time sequential manner, known as frame sequential colour (FSC), or to sequence the bitplanes in a binary modulation device to create a grey scale, or both. Moreover, in the case of a liquid crystal device, it is important to alternate phases of positive and negative polarity in each sub-frame, in order to avoid electrolytic decomposition of the liquid crystal material. These sub-frames are also called the fields. In FSC, a single microdisplay is responsible for modulating the three colours which necessitates a minimum of three sub-frames. When the grey scale depth is constructed in a frame sequential manner, many more sub-frames are required per grey-scale frame. Sub-frame sequential methods rely on the averaging property of the HVS, provided that the frame speed is sufficiently high for this sub-frame averaging. In the case of an EASLM in an optical system, the important parameter for sub-frame averaging is the speed of the camera rather than that of the human eye. The speed at which independent fields can be sequenced on the EASLM is called the refresh rate.

2.3 LIQUID CRYSTAL ON SILICON DEVICES

Liquid crystal on silicon devices were developed initially for the digital watch industry and later for microdisplays [130]. They are all reflective devices except for those where the silicon has been thinned and transferred onto a transmissive substrate such as fused sapphire, for example the SXGA (super extended graphics array) panel from Kopin Corporation (Table 2.1). The large investment in the development of high quality microdisplays has resulted in the availability of high complexity EASLMs for Fourier optical systems. They consist of a liquid crystal layer on top of the CMOS backplane. In order to act as an electrooptic crystal, the liquid crystal layer is aligned using two alignment layers, one on the cover glass and the second on the silicon backplane. The alignment layer defines the optic axis (or director) of the liquid crystal at that interface. A drive cell for the liquid crystal layer is formed by a cover glass with an indium tin oxide (ITO) front electrode, and mirror

TABLE 2.1 Commercial microdisplays suitable for use as EASLMs

Manufacturer	Product	Type	Array size	SBWP	Pixel pitch	Fill Factor	Refresh rate
Canon	WUX450	LCOS	0.71″ diag	1920x1200	8		120 fps
Citizen Miyota	Quad-VGA	FLCOS	0.4″ diag	1280x960	6.35	93.5	120 fps
Compound Photonics	4K	LCOS	0.55″ diag	4096x2160	3.015	94	5.7 kHz
Forth DD	QXGA-R9	FLCOS	0.83″ diag	2048x1536	8.2	96	120 Hz
Hamamatsu	X13138	LCOS	15.9x12.8 mm	1272x1024	12.5	92	360 Hz
Himax	HX7318	LCOS	0.37″ diag	1366x768	6	55	60 Hz
Holoeye	LC 2012	TFT	1.8″ diag	1024x768	36	93	220 Hz
JVC	D-ILA	LCOS	1.27″ diag	4096x2400	6.8	89	24 Hz
JDC	JD8714	LCOS	0.7″ diag	4096x2400	3.74		
Kopin Corp	1280M LV	TLCOS	0.97″ diag	1280x1024	15		
Meadowlark	High Resolution	FLCOS	17.66x10.6 mm	1920x1152	9.2	95.7	868 Hz
Omnivision	OVP2200	LCOS	0.26″ diag	1280x720	4.5		300 Hz
RAONTECH	RDP550F	LCOS	0.55″ diag	2056x1088	6.3		360 Hz
Rockwell Collins	2015HC	LCOS	0.82″ diag	2048x1536	8.1		60 Hz
Santec	SLM100	LCOS	15.0x10.9 mm	1440x1050	10.4	92	
Silicon Micro Display	ST1080	LCOS	0.74″ diag	1920x1080	8.5	92	360 Hz
Sony	4K SXRD	LCOS	0.74″ diag	4096x2160	4		
Syndiant	SYL2271	LCOS	0.37″ diag	1280x720	6.4	93.8	300 Hz
Varitronix	VMD6100	LCOS	0.82″ diag	1920x1280	9	90	60 Hz
Fraunhofer IPMS	1MP SLM	MMA	33x8 mm	2048x512	16	90	2 kHz
Silicon Light Machines	G8192	GLV	43x10 mm	8192x1	5	1	250 kHz
Texas Instruments	DLP9500	DMD	0.95″ diag	1920x1080	10.8	92	7.1 kHz
ViALUX	STAR-07	DMD	0.7″ diag	1024x768			22.7 kHz

pixels on the backplane which can be driven by the CMOS circuitry. The advantage of placing the pixel mirror on top of the circuitry is that the fill factor (FF) of the pixel is high (up to 93%). The FF is the ratio of the area of the pixel mirror to the total area of the pixel, or $ab/\Delta x\Delta y$. This is also known as the aperture ratio (AR). Metal-oxide-semiconductor field-effect transistors (MOSFETs) developed by CMOS processes in the range of 0.18 to 0.35 μm, sometimes with additional high-voltage transistors, provide the necessary voltage to switch the LC. A schematic cross-section of one pixel of the LCOS is shown in Figure 2.3. The LCOS device operates in reflective mode, except for the transmissive type in which the thinned silicon layers have been transferred to transmissive sapphire substrate.

In a reflective LCOS, the incident light is reflected by the top metal layer that has been specially processed for high reflectivity. In the final device the reflectivity is typically between 65 and 80%. The wavefront distortion over the pixellated aperture should be small in view of the use of coherent illumination. Two aspects of the fabrication technology have been developed to achieve this end. Firstly, prior to deposition of the top layer metal, a layer of silicon oxide, for example, is deposited and polished using a combination of chemical and mechanical techniques. This is known as chemical-mechanical polishing (CMP). Secondly, care is taken over the deposition of the aluminium top metal layer. The deposition rate is controlled to give a fine polycrystalline structure to the layer which enhances the reflectivity of the mirror. Moreover, the aluminium is alloyed with copper and annealed at 400°C to prevent the formation of hillocks. In spite of these precautions, the mirror electrode commonly displays a depression at the contact between the vertical via between the interconnect metal layer and the mirror electrode. This is known as the contact divot and is present in most devices.

2.3.1 Configuration of the liquid crystal layer

The selection criteria for the configuration of the LC layer in an LCOS are different from those for a microdisplay. A good overview of the available modes of operation for nematic liquid crystals (NLCs) is given in [165]. The LCOS is used with a given angle of incidence (usually 0°), and has to perform over a narrow viewing angle. It is used at a limited number of wavelengths (usually 1), and the cell gap is maintained to a high precision. For these reasons, a liquid crystal configuration known

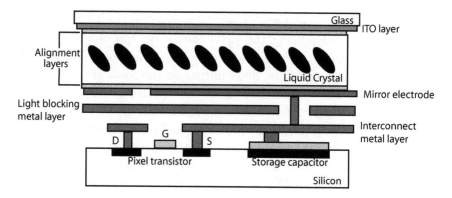

Figure 2.3 Cross-section of pixel in LCOS device

as the planar aligned nematic (PAN) cell, which is no longer used for microdisplay because of poor speed and uniformity, is attractive for both amplitude and phase modulation. An alternative name for this effect is the zero degree twist or electrically controlled birefringence (ECB) cell. The birefringence modulation consequent on applying a voltage to the cell generates a phase modulation effect when the polarisation of the light is aligned with the liquid crystal director, as used in the Hamamatsu X13138 and JDC JD8714 devices. A phase modulation of 2π can be achieved in a cell where the optical path difference between the light reflected from a pixel in the zero voltage state and one in the fully driven state is one wavelength. If it is desired to use this effect for amplitude modulation, the polarisation of the light is aligned at 45° to the liquid crystal director, and the thickness of the LC layer is halved because the optical path difference required is only one half wavelength for full amplitude modulation.

A second liquid crystal effect called the vertically aligned nematic (VAN)is used predominantly in microdisplay because of the excellent contrast between the black zero voltage state and the driven state. This is also known as the tilted homeotropic (TH) mode. It is used in the JVC LCOS for projector applications. The liquid crystal used in such devices has a negative dielectric anisotropy where the long axis of the molecule aligns perpendicular to the applied field. The availability of good liquid crystal mixtures for this mode is more limited than the more common positive dielectric anisotropy liquid crystals used in the other modes. However, it has potential use in EASLMs if high birefringence liquid crystals can be developed.

The liquid crystal configuration of the two modes which have been discussed does not twist within the cell. Therefore, when the electric field driving the pixel is removed, the liquid crystal is slow to return to the equilbrium state due to hydrodynamic flow in the liquid crystal medium. Twisted liquid crystal modes benefit from a reduced flow and the liquid crystal returns to its equilibrium state more quickly. Three modes which employ twist are the 45° twist mode (also known as the hybrid-field-effect (HFE) mode), the twisted nematic (TN) mode, and the mixed twisted nematic (MTN) mode. The HFE mode has a 45° twist and an excellent tolerance to cell gap non-uniformity. This was important in the early days of optically addressed spatial light modulators (OASLMs) (Section 2.6) where the backplane had poor planarity [23]. The TN mode is a 90° twist cell which is used in transmissive devices. The polarisation vector of the incident light is aligned with the director of the liquid crystal cell on the front surface of the cell in both the HFE and TN modes. The MTN mode mixes the polarisation rotation property of a TN mode with the birefringence modulation effect of the PAN mode. In common with the TN mode, there is a 90° twist, but the incident polarisation for reflective devices is at 20° to the alignment of the front surface. The 90° twist produces a fast switch speed in both the TN and MTN devices. Further speed improvement can be achieved with the optically compensated bend (OCB) mode. This can be viewed as a 180° twist cell because the alignment pre-tilt on the opposing substrates of the cell is in the opposite sense (Figure 2.4). The idea for an OCB LCOS was presented in [279].

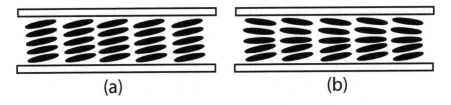

(a)   (b)

Figure 2.4 Cross-section of (a) PAN cell; and (b) OCB cell, illustrating opposite sense of alignment pre-tilt in the latter

A minority of LCOS devices employ ferroelectric liquid crystals (FLCs) and these are denoted in Table 2.1 by the acronym FLCOS. FLCs are a more ordered phase of liquid crystals than the NLCs, known as chiral smectic C phase. The suitability of using FLCs in DRAM-type CMOS backplanes was first addressed in [61]. The configuration of the director in these devices possesses two stable in-plane orientations. The liquid crystal is switched between these two orientations using the polarity of the electric field generated by an SRAM backplane. This is completely different from the case of nematic LCOS, where the magnitude of the rms voltage is used. In the FLCOS, the liquid crystal is driven into both of these two orientations resulting in a fast frame rate. The angle between the two orientations of the director is determined by the chemistry of the liquid crystal mixture employed, but a value between 20° and 40° is common. Fast binary modulation of amplitude, phase, and polarisation can be achieved in this device. This ferroelectric technology was finding its way into devices at around the same time as demonstrator correlator systems were being constructed (Chapter 6). A multiplexed FLC device, where passive matrix address technology was employed, was used in one of these systems. The speed of passive matrix devices is low because the rows of the matrix must be addressed "line-at-a-time"; the SBWP is also low because the pixel spacing and active area are limited by photolithography. The multiplexed FLC device in Table 6.1 is given the acronym FLCD. A summary of this and other modes of modulation in smectic liquid crystals is given in [227].

2.4 MICRO-OPTO-ELECTRO-MECHANICAL SYSTEMS

The bulk micromachining of a silicon wafer to form membrane structures was developed in the 1960s and has been progressively refined since then. Deformable-mirror devices based on bulk micromachining have been under development since the late 1970's [104]. They were the first example of a micro-opto-electro-mechanical system (MOEMS) (also known as optical MEMS (micro-electro-mechanical systems)), which were originally targetted at analogue and digital modulation of both amplitude-dominant and phase-dominant modes of modulation. In 1987, the digital micromirror device (DMD) was launched, and the Texas Instrument's (TI) digital light processing (DLP) technology is based on the DMD microchip. The pixel mirrors rotate about a 45° torsion beam. The 'on' and 'off' state positions of the mirror are separated by 24°. Each mirror is electrostatically turned on or off using pulse width modulation, with a modulation frequency more than 10 kHz. Each rectangular mirror generates a *sinc* diffraction pattern (Figure 2.1). Since the mirror deflects the beam along the 45° diagonal, there is a spatial separation between the deflected beams and the subsidiary maxima of the sinc pattern giving an excellent extinction ratio from 'on' to 'off'.

The DMD microchip is the most common form of optical MEMS. There are other types which are more suited to Fourier optics, such as the analogue micromirror arrays (MMAs) and piston-type MEMS. The Fraunhofer IPMS Institute has fabricated a 1 megapixel MMA, where the tilt of the mirrors can be controlled with an addressing voltage of 10 bit resolution, where 'resolution' here refers to the granularity of the addressing voltage. The amplitude of the retro-reflected beam decreases from 100% reflection to zero in a smooth fashion as the mirrors are tilted. Piston-type MEMS have also been made by the same establishment, where they are known as piston MMA. Piston is the mean value of the phase profile across the pixel aperture. The pixel is actuated up and down, as in a spring balance. The maximum deformation is known as the stroke, which is 0.4 microns in the Fraunhofer device [150]. This is adequate for Fourier optics application. The descendent of the original deformable-mirror device, which is now known as the deformable mirror (DM) is not suited for the Fourier optics systems described in this book.

The surface micromachining of a silicon wafer is the basis of the grating light valve (GLV) technology. A layer of silicon oxide followed by a layer of silicon nitride is deposited on the silicon wafer. The silicon nitride is patterned into narrow strips, followed by the etching of the silicon oxide layer. Finally, aluminium is evaporated onto the active area of the device. The etching of the silicon oxide releases the silicon nitride ribbons from the silicon substrate so that an applied voltage can move them down towards the substrate. The thickness of the oxide layer is designed to be one quarter of the wavelength of light which is to be used, so that, when the ribbon is pulled down onto the substrate, the path length of the light reflected from the ribbon is changed by one half wavelength. The interference between light reflected by the ribbon and light reflected from the substrate mirrors, in between the ribbons, changes from one of constructive interference, when no voltage is applied beneath the ribbon, to one of destructive interference, when a voltage is applied. In the state of destructive interference, the light is diffracted at an angle given by the spatial frequency of the grating multiplied by the wavelength of the light.

In both types of optical MEMS, the incident light is deflected away from the normal by applying a voltage to the pixel. In the DMD, this is reflected on a mirror and, in the GLV, it is diffracted. The GLV technology is more suited to a 1D EASLM. The available modes of operation of the DMD are more limited than with the LCOS devices, but the operation speed is faster. It acts as a binary amplitude modulator, and the light is either deflected or not. The GLV gives a pure analogue modulation, and therefore acts in a similar manner to the phase-modulating ECB LCOS. There is no polarisation rotation and a high reflectivity.

2.5 TRANSMISSIVE LIQUID CRYSTAL DEVICES

Two types of transmissive liquid crystal devices are available: the thinned LCOS Kopin device (Table 2.1) and TFT screens removed from commercial projectors. The latter formed an important part of early correlator systems (Section 6.7). Both types of device employ a liquid crystal layer in the TN mode of operation. The devices are conventionally used for amplitude modulation with an analyzer at the rear of the cell. If the analyzer is oriented parallel to the polarisation of the incident light, the 'off' state is low transmission and this is referred to as normally black (NB) operation. NB operation can provide high contrast ratios. If the analyzer is oriented perpendicular to the polarisation of the incident light, the 'off' state is high transmission and this is referred to as normally white (NW) operation.

The TN mode is used for mixed amplitude/phase modulation along both axes of the electric field vector. In the 'off' state, the output electric field vector is perpendicular to the input electric field vector. This is a result of the twist in the cell rotating the polarization vector. As the voltage across the cell is increased, the liquid crystal molecules at the centre of the cell begin to tilt towards the bounding surfaces of the cell. The result is that the polarization vector at the output is now elliptical, due to a component parallel to the incident polarization which is advanced in phase compared with the rotated polarization component. Further increase in the voltage leads eventually to the destruction of the twist and the light becomes linearly polarized along the direction of the incident light vector.

2.6 OPTICALLY ADDRESSED SPATIAL LIGHT MODULATOR

In the late 1960s and 1970s there was a large effort in numerous laboratories to fabricate optically addressed spatial light modulators. These devices allow the image to be written directly into the optical system, via an unpixellated photosensor. The photosensor activates a proximate light modulating material based on one of the following types of modulator: electro-optic crystals; ferroelectric ceramics; deformable elastomers, liquids, and gels; deformable membrane devices; and liquid crystals [153]. The most successful material has been the liquid crystals due to the facility of device construction and the small voltages required for the electrooptic effect. In order to parameterise the SBWP of these unpixellated devices, the resolution in linepairs per mm (or lp/mm) is used. In this context, the resolution refers to the maximum spatial frequency which can be recorded on the photosensor and reproduced by the light modulating material. The fidelity of image transfer between these two layers is determined by a number of device factors which are discussed in, for example, [272]. The resolution of the device is usually quoted at 50

The liquid crystal light valve (LCLV) is a sandwich of the liquid crystal with a photosensor layer. (Certain authors use this term to refer to the electrically addressed devices discussed in Sections 2.3 and 2.5.) LCLVs do not require a complex drive electronics, a fact which simplifies their fabrication. A single voltage applied across the LCLV results in a voltage division between the photosensor and the liquid crystal dependent on the light pattern on the photosensor. The task of the device design engineer is to ensure that the voltage is dropped across the photosensor in the areas which are not illuminated, on the one hand; and across the liquid crystal in the illuminated areas, on the other hand. Many examples of this structure have been researched. The resolution can be measured by a variety of methods [45]. The focus here is on devices which have been sold commercially, and incorporated into Fourier optical systems.

The Hughes LCLV was announced in 1975 as a non-coherent to coherent image converter [92]. The bottleneck in coherent optical data processing systems was perceived to be the use of photographic images on the input to the system. The Hughes device could not be operated in real time, because the frame rate was between 5 and 14 Hz [112]. An active aperture of 1 square inch, and a resolution of 20 lp/mm (at 50% MTF) offered an input SBWP in excess of 10^5. The photosensor layer was CdS and the liquid crystal layer was the mixture E7 from Merck aligned in the HFE mode. In order to operate the device in reflection, a dielectric mirror was interposed between the liquid crystal layer and the CdS. The coherent (read) beam was incident on the LC layer and reflected by the dielectric mirror. The incoherent (write) beam was incident on the CdS and absorbed in this layer. In order to increase the isolation between read and write beams, a CdTe light-absorbing layer was sandwiched between the CdS layer and the dielectric mirror. At the same time, parallel efforts at making the LCLV were pursued in both Europe and Russia. The European venture employed a CdSe photoconductor [133], and the Russian works employed GaAs [141] and chalcogenide glass [204].

The PAL-SLM (X5641) was fabricated by Hamamatsu Photonics K.K. CRL between 1992 and 2006. It replaced the microchannel OASLM which the company had been making since 1985 [99]. This was a LCLV based on amorphous silicon photoconductor. In contrast to the Hughes LCLV, the liquid crystal layer was a PAN cell and provided greater than 2π phase modulation at red wavelengths, where the frame rate was around 30 Hz. An active aperture of 18 x 18 mm^2, and a resolution of 50 lp/mm (at 50% MTF) offered a superior input SBWP to the Hughes LCLV. Although the PAN cell alignment was used rather than the HFE twisted mode, the frame rate was superior to that of the Hughes LCLV because the frame rate in these devices is determined by the time constant of the photoconductor/liquid crystal structure. The amorphous silicon layer improves the time constant in the LCLV in comparison with the CdS layer. The company also made a TN and an FLC LCLV, X4171 and X4601, respectively. The resolution and response time of the TN device were similar to the PAL-SLM. However, the X4601 offered a higher resolution (100 lp/mm) and shorter response time (100 μs combined on and off times). Hamamatsu has discontinued production of all these devices.

An LCLV using photoconductive bismuth silicon oxide (BSO) as the photoconductor was fabricated by the Central Research Laboratory of Thomson-CSF in Orsay, France [12]. Although the resolution was only 12 lp/mm, this photosensor allowed transmission mode operation when a green or blue write light was used in conjunction with a red read light. Large area, thin crystals of BSO can be purchased from Sillenites Ltd of St Petersburg for the purpose of making BSO LCLVs with apertures up to 40 x 40 mm^2. More recently, the transmissive LCLV has also been fabricated using thin chalcogenide layers [138], thin zinc oxide [236], and gallium nitride [228], as the photosensing layer. Transmissive devices can be useful for Fourier optical systems in simplifying system design. The reflective LCLVs remain useful as input transducers to the optical systems: capturing a scene on the photoreceptor and transfering this into the coherent optical system by means of the liquid crystal layer. The speed of devices which employ nematic liquid crystals is inferior to the electrically addressed nematic devices. Higher speed reflective devices based on ferroelectric liquid crystals have been extensively researched [181]. CRL Smectic Technology marketed a ferroelectric liquid crystal OASLM called the BOASLM (Binary OASLM) in the early 1990s. It was based on a hydrogenated amorphous silicon photoconductor and had a write time of between 100 and 500 μs, with a resolution of greater than 50 lp/mm (10% MTF). The write sensitivity was less than 30 $\mu W/cm^2$ and the active area was a 30 mm diameter circle.

The OASLMs used in the systems which will be described in Chapter 6 are the LCLVs. Therefore, EASLMs will be referred to as SLMs in the rest of the book, retaining LCLV for the OASLMs which have been used.

GLOSSARY

Apodisation: The word literally means "removal of the feet". It is used to refer to the removal of the light amplitude by a hard aperture, such as that outside of the reflecting pixellated surface of an EASLM, or a soft aperture, such as, for example, that of the *sinc* function.

Liquid crystal light valve (LCLV): A hybrid device incorporating a photoreceptor layer and a liquid crystal layer. The photoreceptor captures a weak image signal which can be transferred to a more powerful light beam by the liquid crystal layer.

Ronchi grating: A constant interval mark and space square wave optical target with unity mark/space ratio.

Space bandwidth product (SBWP): The total number of pixels in an SLM. This can also be applied to the number of spots in the focal plane of a lens, or the number of data streams in an optical processing system.

Spatial light modulator (SLM): An electrooptic device which modulates spatially a light beam transmitted or reflected by the device.

CHAPTER 3

Diffractive optical elements

CONTENTS

3.1   INTRODUCTION

In order to transform and filter the images within an optical processing system, a useful tool is the diffractive optical element (DOE). The DOE is a beam shaping optical element for a coherent light beam, where beam shaping is achieved by significant modulation of the refractive index of the element on the microscale. In contrast to refractive optical elements such as a lens, or a prism, the modulation of refractive index occurs over shorter spatial distances. Due to this high spatial frequency modulation, individual elements of the image can be processed differently to other elements, so that geometrical transformation of the image can be accomplished [34]. This is known as space variant interconnection in contrast to the space invariant interconnection provided by a lens or a bulk prism. DOEs are widely used in optical systems, which require space variant connections, such as optical interconnects and multichannel Fourier optical systems. Feature extraction and image transforms with DOEs have been described in [222, 224, 225, 226].

The first hand-drawn diffraction element was a matched filter for an optical correlator [33]. This was called a detour phase hologram because the phase information was encoded in the position of slits in each amplitude modulated pixel. The encoding was effected by decentring the slit by an amount corresponding to the magnitude of the phase.

The phrase "computer generated hologram" (CGH) was used by the same group to describe the use of a computer-guided plotter to create the detour phase holograms [162]. This development allowed the mathematical definition of wavefronts by structures plotted onto paper and subsequently demagnified onto photographic emulsion by means of a photoplotter. The highest resolution of these binary structures was defined by a cell size of 10 micron. Gray scale in amplitude was achieved by the size of the aperture in an opaque cell, and in phase by detour phase coding. This allowed quantization of both the amplitude and the phase to around 15 levels. A survey of CGH techniques for coding complex amplitude with an extensive bibliography is given in [260].

The resolution of the diffraction element has subsequently been increased due to the development of specialised machines for lithography and surface micromachining. The diffraction element is a surface relief structure and a multi-level structure is achieved by exposing the substrate to a sequence of binary masks of decreasing feature size. Due to the precise positioning available in these machines, gray scale based on multilevel quantization can be accurately controlled. This field was given the name "binary optics" [76]. By anchoring the fabrication technology in lithographic processes, the resolution of the devices and elements scales with the resolution of lithographic processes. The latter are pushed towards higher resolution by the semiconductor industry's desire to maintain Moore's law progression with integrated circuit technology. Therefore, the resolution of binary optics will improve in a similar manner. Similar arguments apply to the SLMs based on integrated circuit technology, with the proviso that the masks used for binary optics are generally of higher resolution than those used for device fabrication.

The general term, DOE, comprises both CGHs and binary optics, and is the term generally used nowadays. The design and modelling of DOEs is covered in dedicated books, such as [144]. An introduction to the design of these elements is given in Section 3.2. Subsequent to the design of these elements, an encoding scheme must be employed in order to convert the design to a data file which is suitable for the mask maker. Popular schemes will be outlined in Section 3.3. The fabrication possibilities will be listed in Section 3.4. Some of the fabrication houses will perform the design also, and these will be indicated.

3.2 DESIGN OF DIFFRACTIVE OPTICAL ELEMENTS

The first CGH was an amplitude mask. Light absorption restricts the efficiency of such structures. If we select the modulation of the phase of the wavefront in place of the amplitude, then a much more efficient wavefront modulation is achieved. The kinoform was developed to fulfil this promise of a higher efficiency [156]. In addition, an algorithm was presented which reduced the time for computing the structure. The design of DOEs is nowadays targeted on pure phase structures.

The main methods for design of DOEs are analytic techniques, scalar diffraction, and rigorous design based on electromagnetic theory. Analytic techniques can be applied in the design of DOEs for geometrical transformations between two planes [34]. They can lead to complexities in the detail of the hologram which can be avoided when an equivalent refractive element is used. For example, the phase singularities which appear in a DOE for 90-degree rotation of an image [223] can be avoided by the use of a Dove prism to effect the rotation. The application of a Dove prism to a Fourier optical system is explored in [100]. A common transformation for Fourier optical systems is the Gaussian to flat top beam intensity converter, which converts the Gaussian intensity profile of the incident laser beam to a uniform intensity profile. Both refractive [103] and diffractive [267] solutions have been been introduced and explored further in later works. In view of the rectangular profile of an EASLM, an additional shaping from a circular top-hat to a rectangular top-hat improves the efficiency of both solutions. Apart from the use of birefringent media [151], this circular to top-hat conversion necessitates the use of DOEs.

Scalar diffraction theory is applied to the calculation of the majority of DOEs. Under the paraxial approximation, Equation 1.6, the transform can be inverted in order to derive the aperture function from the required output plane. In order to constrain the aperture function to the requirements of the DOE, an iterative algorithm can be employed (Figure 3.1).

This starts from a complex amplitude version of the desired (target) diffraction pattern in the output plane (labelled Input in Figure 3.1). The initial complex amplitude target is, typically, the product of the amplitude target pattern and a random complex phasor or a quadratic phase distribution. This is transformed to the DOE plane using the inverse Fourier transform (IFT) (Equation 1.9). The DOE constraint, e.g., constant amplitude for a phase-only DOE, is applied to the result

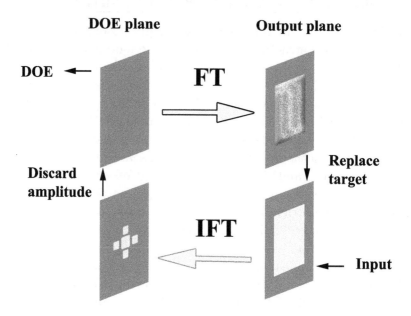

Figure 3.1 Graphical representation of the iterative Fourier transform algorithm

of the inverse Fourier transformation (labelled Discard amplitude in Figure 3.1). This is followed by a forward FT to return to the output plane. If the difference between the result of the forward FT and the true target is greater than a value set by the designer, then the amplitude distribution in the output plane is replaced by the amplitude of the target (labelled Replace target in Figure 3.1). When this cycle of inverse and forward FT is repeated, it is found that the difference between the result of the forward FT and the true target gradually diminishes with increasing number of cycles [89]. When the difference is less than a value set by the designer, the resulting DOE is output (top left of Figure 3.1). A suitable criterion is the mean squared error (MSE), where the square of the difference between the output image and the target is computed on a pixel-by-pixel basis and summed. The initial description of this algorithm solved the phase of a 32 x 32 pixel image in under 80 seconds [89]. In this description, thirty seven FTs reduced the mean squared error between the result and the true target from 73.0 to 0.01. With current digital processing power, the processing time for this algorithm has been reduced by three orders of magnitude.

The iterative algorithm has been used for a number of design problems, including beam shaping [266]. In particular design problems, the original algorithm described by Gerchberg and Saxton [89] stagnates. Stagnation is the inability to reduce the error below a floor which exceeds the desired error criterion. A large number of techniques have been developed for avoiding stagnation in particular instances, but no standardised procedure has emerged that is suitable for a wide range of problems. A significant improvement was the development of three more refined algorithms, the input-output [77], the output-output, and the hybrid input-output algorithms [78]. The choice of one or other of these algorithms is the prerogative of the designer. The totality of these with other algorithms developed since are known as IFTAs (Iterative Fourier Transform Algorithms). A significant improvement has been the introduction of a noise dump area surrounding the DOE [88]. A further improvement is the introduction of additional terms into the MSE criterion with adjustable parameters. These allow the DOE designer to control the minimisation rate of the new criterion and partially eliminate stagnation [143]. Improvements in this methodology allow the optimisation of the uniformity and efficiency of the DOE at the same time [134].

The result of the optimization algorithm is a design for the DOE which is not unique. If the algorithm is repeated with a different random phase seed, then a different design will result. The performance (efficiency and MSE) of different designs will be comparable but distinct. A more computationally expensive design method which does not stagnate is simulated annealing (SA) [136]. Starting from a random phase distribution, the phase hologram is progressively refined by making changes to each pixel, checking the MSE, and accepting the change or not, depending on the "throw of a dice", made at each change. This algorithm is considered to give the best solution to the design problem. The MSE is known as the cost function which can be further elaborated according to any additional constraints required in the optimisation, for example the efficiency.

The Gerchberg-Saxton algorithm generates a phase distribution which, under illumination with a collimated coherent light beam, will reproduce the target pattern in the far field. If it is desired that the pattern is generated in the near field, then the transforms applied in the iteration loop must be adapted accordingly. Fractional Fourier and Fresnel transforms were applied to this problem in [287]. Neither of these transforms possesses an inverse, so the propagation from the target plane to the DOE plane is achieved by negating the fractional order and negating the propagation distance, respectively, in the two cases. An alternative technique is to use an iterative algorithm based on the angular spectrum method [179]. The angular spectrum method, also known as the plane wave spectrum (PWS) method, is detailed in [90] and [212]. The elementary wavelets in the Rayleigh-Sommerfeld integral, Equation (1.2), are expressed as a plane wave expansion by the formula developed in [281]. The propagation of plane waves between parallel planes is more readily computed than the propagation of spherical waves, using the formula

$$U(x, y, z) = \int\limits_{-\infty}^{\infty}\!\!\int \hat{U}(f_x, f_y) exp[ik\sqrt{(1 - (f_x^2 + f_y^2)\lambda^2)}z].$$

$$exp[2\pi i(f_x x + f_y y)] \, df_x df_y \tag{3.1}$$

where z defines the optic axis and x, and y defines position in a plane perpendicular to this axis. When $(f_x^2 + f_y^2)\lambda^2 \ll 1$, this can be simplified to

$$U(x, y, z) = exp[ikz] \int\limits_{-\infty}^{\infty}\!\!\int \hat{U}(f_x, f_y) exp[-i\pi\lambda z(f_x^2 + f_y^2)].$$

$$exp[2\pi i(f_x x + f_y y)] \, df_x df_y \tag{3.2}$$

The use of the PWS formula, Equation (3.2), results in a reduction in computational complexity from N^2 to $NlogN$ [194].

Evanescent waves

The plane wave waves which propagate to the output plane satisfy $(f_x^2 + f_y^2)\lambda^2 \leq 1$. If $(f_x^2 + f_y^2)\lambda^2 > 1$, then the wave amplitude diminishes as it propagates along the z axis. This is an evanescent wave. These waves do not contribute to the output plane, and therefore represent energy loss in the transfer from the DOE to the output plane. They can be neglected in scalar diffraction designs, but are more important in rigorous design where the feature size is smaller and spatial frequencies are higher.

The rigorous design of DOEs is required when the feature size of the DOE is reduced to the order of magnitude of the wavelength of the illuminating light beam. This is currently a research field, although there are a number of application areas, such as wire grid polarizers. High resolution DOEs could also be used in Fourier optical systems when the size of the system shrinks or when it is desired to construct the system in solid optics and high resolution gratings are required. The mathematical techniques for solving Maxwell's equations have been researched under a variety of names, such as Rigorous Coupled Wave Analysis (RCWA), Fourier Expansion Modal Methods (FMM), Finite Element Methods (FEM), and Finite Difference Time Domain (FDTD) techniques. The first two, RCWA and FMM, are synonymous, and cover the mathematical techniques used to expand the electromagnetic field as a Fourier series for which the propagation through the DOE can be computed relatively easily. Commercial software packages for RCWA include GSolver and DiffractMOD (Rsoft). FEM divide the DOE into a mesh of elements, commonly triangular, where the electromagnetic field is expressed as a linear combination of elementary basis functions. The conjunction of these linear approximations over the constellation of elements and the matching with the boundary constraints form the essentials of the method. JCMwave is an example of an FEM solver that can be used for DOE design. FDTD is a numerical solver based on finite-difference approximations to the derivative operators in Maxwell's differential equations. A number of FDTD solvers are available, for example that of Lumerical Inc.

There are software packages, such as VirtualLab, which cover most of the design techniques presented above, but a large number of research workers prefer to develop their own algorithms. The output of the design is a wavefront which will generate the desired diffraction pattern. This is expressed as a complex amplitude matrix if a digital computer is used. Then this design has to be written onto a physical mask. Direct coding of the complex amplitude matrix, such as detour phase, is no longer used, due to its relative inefficiency. The invention of the kinoform [156] unveiled both a design procedure and the efficiency advantages of coding the DOE in a phase relief structure. Multi-level phase DOEs have a higher efficiency than simple binary phase DOEs. The choices for fabricating a phase relief structure will be presented in the next section.

3.3 FABRICATION

In respect of the construction of optical systems, the optical designer is the poor cousin of the electronics design engineer. The availability of a variety of off-the-shelf components is nonexistent. This is no more evident than for DOEs. The situation is compounded because the organisations with the capability for the fabrication of these elements will not necessarily furnish one-off elements for optical engineers prototyping new systems. Two exceptions are Holo-or and Jenoptik, which will fabricate a phase-only element from a custom design. Alternatively, a larger number of companies, such as Compugraphics, JD Photo Data, Delta Mask, and Toppan will take an in-house design and convert it into an amplitude-modulating photomask.

A photomask is a high-resolution pattern of aluminium/chrome on a soda lime or quartz substrate. This amplitude mask should be designed for eventual production of a phase DOE, because of the much higher efficiency of the phase DOE. The phase DOE design, as detailed in Section 3.2, is a matrix of real numbers. This matrix has to be dimensioned, so that the physical extent of each element is defined, with reference to the optical system in which it will be inserted. The quantized DOE design is converted to a series of mask designs using a layout editor. This mask design can then be converted to a binary phase element in a single-exposure lithography system, such as a wafer stepper or step-and-scan system. The substrate for these lithography systems is the material which is to be etched, such as glass or silicon, covered by a thin layer of high contrast photoresist. In order to expose the photoresist, the substrate is placed in the lithography system together with the mask. The system allows a choice of either contact, proximity, or projection exposure. Contact exposure gives the highest resolution but results in debris from the photoresist contaminating the mask. Unfortunately, the masks cannot be repeatedly cleaned without some degradation of the DOE, therefore proximity is generally preferred. After exposure, the substrate is removed from the lithography system and immersed in a developer solution. If a positive tone resist has been employed, the areas which have been exposed to the light in the lithography system are removed by the developer. If a negative tone resist has been employed, the areas which have been protected from light by the aluminium/chrome areas of the mask are removed by the developer. Subsequent to the development of the photoresist, the design is etched into the glass or silicon substrate using, for example, an ion beam etch lithography , sometimes called an ion milling machine. The areas where resist remains on the substrate are somewhat protected from the etching in this machine so that the mask pattern is transferred into the substrate. A faster etch rate can be achieved in a reactive ion etch (RIE) lithography, although impurities in the glass substrate may lead to an uneven etch. The etch depth in both types of etching systems must be calibrated for a specific design, usually by producing several runs in order to meet the target etch depth. Finally, any residual resist on the substrate is removed. In order to realise a multilevel phase element, the substrate is further coated with photoresist and exposed with a higher resolution mask. This procedure is repeated using multiple exposures with successively higher resolution masks [76].

An alternative to the photomask is a laser lithography system, where analogue photoresist phase elements can be made in a one-step process. Both Mycronic AB and Heidelberg Instruments supply laser lithography systems. The high-end systems of both companies can write to a 10 nm grid with a minimum feature size of 0.75 and 0.6 microns, respectively. Other companies, such as RPC Photonics, use homemade laser lithography systems, and the Institute of Automation and Electrometry use a circular laser writing system [205]. The photoresist used in greyscale mask laser lithography is either a thin layer of low contrast resist, or a thick layer ($> 5 \ \mu m$) of resist. Due to the bleaching of the resist layer during the UV exposure, the exposure depth of the thick layer is proportional to the exposure dose. These analogue photoresist phase elements can be used with red illumination. In order to produce a greyscale mask which is permanent and which will not be affected by subsequent ambient UV illumination, the pattern can be etched into the substrate using, for example, an ion beam etch. If the etch preserves the linear differentiation between exposed and unexposed areas, then the greyscale mask can be faithfully imprinted in the substrate.

The feature size of a step-and-scan system is much smaller than that of laser lithography systems, e.g., sub 100 nm, but access to these state-of-the-art systems is correspondingly more restricted. For smaller feature sizes, e.g., 20 nm, electron beam lithography is used, using an e-beam resist in place of the photoresist layer used in laser lithography. This is the lithography used, albeit at larger feature size, in the generation of the photomask. For optimal results, e-beam lithography is the preferred lithography system [11]. Fabrication constraints can also be incorporated into the design process [191]. E-beam lithography is also used for the generation of tooling for nanoimprint lithography. In conjunction with high-resolution deposition processes, such as atomic layer deposition, high-aspect-ratio structures can be realised [132]. DIOP-

TIC GmbH offer RIE of the DOEs that they have produced by either laser or e-beam lithography. Raith GmbH sells equipment for e-beam lithography.

In semiconductor lithography systems, a considerable effort is spent on eliminating the errors which arise during fabrication of high-resolution structures over relatively large areas. Potential error sources arise due to pattern distortion, misalignment, and substrate distortion. Pattern distortion will be minimal when the address grid is finer than the feature size. Misalignment is a significant problem at the multiple exposure step for multilevel phase elements. Substrate distortion is usually accommodated by replication onto higher optical quality substrates. In order to reduce potential errors during the design phase, the SA algorithm can be constrained to favour contiguous areas of similar phase level in the DOE.

In order to realise low-cost optical systems, it is important to be able to replicate the DOEs. A variety of embossing, moulding and casting techniques have been developed for optical elements. Embossing is the process of pushing a seal into hot wax. A silicon or quartz seal is called a stamp. Copies of silicon or quartz stamps can be made in nickel, and the resulting elements are called nickel shims. Companies, such as NIL Technology, and EV Group make stamps and shims for nanoimprint lithography. CDA GmbH have teamed up with Holoeye Photonics AG to offer a design and fabrication competence for embossing DOEs into either Polymethyl Methacrylate (PMMA), Polycarbonate (PC), Cyclic Olefin Copolymer (COC), or Cyclic Olefin Polymer (COP). Moulding describes the process of using fluid materials to fill the space between the mould and a reference surface, so that, when cool, the injected material takes the form of the mould. Nickel shims, fabricated from the lithographically mastered DOE, are used for injected moulding [192]. Companies, such as NIL Technology, Temicon, and Rothtec will fabricate shims to customer specifications. Casting is a higher quality process than either injection moulding or embossing. The UV-curable or sol-gel materials are introduced between the mould and a substrate, and subsequently cured by either UV light or thermal curing. The replica formed by one of the above techniques is now a low-cost component for the optical system, provided that the quantity required is sufficient to justify the expense of replication.

GLOSSARY

Binary optics: The use of VLSI fabrication techniques such as photolithography and etching, to create high-resolution surface-relief structures in optical substrates.

Detour phase coding: In the context of a complex hologram where the amplitude of the wave transmitted through an opaque mask is defined by the size of the aperture at each pixel, the phase is encoded by the position of the aperture within the pixel.

Space invariant interconnect: An optical interconnect between an input and output plane, where each part of the input is connected in a similar manner to the output plane.

Space variant interconnect: An optical interconnect between an input and output plane, where each part of the input is connected in a dissimilar manner to the output plane.

CHAPTER 4

Transfer functions and cameras

CONTENTS

4.1 INTRODUCTION

A camera is composed of an image sensor with an associated optical element. In consumer cameras, the optical element is an imaging lens system, which images an object onto the image plane where the image sensor is located. Cameras are used for image capture for the input to Fourier optical systems. Image sensors are used at the output plane and on intermediate image planes within the Fourier optical system. Significant data reduction within the system suggests that high-complexity image sensors may not be optimal at the output of the system. In order to understand the limitations of the imaging process, the optical transfer function will be described in Section 4.2. This is followed by an overview of image sensor technology, with a list of exemplar products in the marketplace.

4.2 TRANSFER FUNCTIONS

The imaging process of an optical system is modeled as a linear system. Therefore, if an input, $I(x_i, y_i)$, produces an output, $O(x_o, y_o)$, then a linear sum of inputs produces a linear sum of outputs with the same linear coefficients, i.e.,

$$I(x_i, y_i) \longmapsto O(x_o, y_o)$$

$$aI_1(x_i, y_i) + bI_2(x_i, y_i) \longmapsto aO_1(x_o, y_o) + bO_2(x_o, y_o) \qquad (4.1)$$

where $O_1, O_2(x_o, y_o)$ are the outputs corresponding to inputs $I_1, I_2(x_i, y_i)$, and a, b are constants.

An important consequence of this linearity is that the output can be written as a summation of the inputs each multiplied by a constant which is a function of the coordinates of the input, output and the wavelength of the light, λ,

$$O(x_o, y_o) = \int\!\!\!\int_{-\infty}^{+\infty} W(x_o, y_o; x_i, y_i; \lambda) I(x_i, y_i) dx_i dy_i, \qquad (4.2)$$

where $W(x_o, y_o; x_i, y_i; \lambda)$ is the optical spread function [160].

A second aspect of the modelling is the assumption that the linear system is spatially invariant, so that W depends only on the difference in coordinates $(x_o - x_i)$ and $(y_o - y_i)$ and the wavelength of the light

$$W = W_A(x_o - x_i, y_o - y_i; \lambda) \qquad (4.3)$$

where the subscript A refers to the area or patch over which the system is spatially invariant. The condition of spatial invariance is known as the isoplanatic condition, and the patch is known as the isoplanatic patch. Although the isoplanatic condition applies over a limited area only, the integral equation is used with the proviso that W varies according to location in the input/output planes.

When the input is a monochromatic point source, the spread function is known as the point spread function (PSF), $h(x_i - x_o, y_i - y_o)$ [90] and

$$O(x_o, y_o) = \int\!\!\!\int_{-\infty}^{+\infty} h(x_o - x_i, y_o - y_i) I(x_i, y_i) dx_i dy_i \qquad (4.4)$$

In Equation (4.4), the integral is a convolution of the input with the point spread function. In Section 1.3.3, it was shown that a convolution integral can be written as the product of two FTs

$$\hat{O}(f_x, f_y) = H(f_x, f_y) \hat{I}(f_x, f_y) \qquad (4.5)$$

where $\hat{O}(f_x, f_y)$ is the FT of $O(x_o, y_o)$, $\hat{I}(f_x, f_y)$ is the FT of $I(x_i, y_i)$, and $H(f_x, f_y)$ is the FT of $h(x_o - x_i, y_o - y_i)$. $H(f_x, f_y)$ is called the optical transfer function (OTF).

The transfer function approach is valid for an ideal isoplanatic optical system. This ideal system is one where the translation of an object point in the object plane produces a proportional translation of the image point in the image plane. The diffraction-limited OTF is the transfer function for an ideal imaging system which is free from aberrations. The shape of the diffraction-limited OTF as a function of frequency is determined by the limiting aperture of the system, for example, the circular frame of the lens. This has been calculated for both square and circular limiting apertures [90]. It decreases monotonically from a maximum at zero spatial frequency to zero at a cut-off frequency, which is equal to the reciprocal of the product of the f-number of the lens and the wavelength of the light. The f-number is the focal length of the lens divided by the lens aperture. Higher f-number lenses have lower cut-off frequencies.

The diffraction-limited OTF is an important metric for evaluating the performance of camera lenses. The OTF of camera lenses is, in general, a complex quantity, since it includes the aberration of the lens. Therefore, in order to compare it with the diffraction-limited OTF of the limiting aperture of the imaging system, the modulus of the OTF of the lens is used. This is called the modulation transfer function (MTF). The MTF is measured using groups of Ronchi gratings, each grating of a different spatial frequency. When the lens is used to make a unity magnification image of these gratings, the contrast in the image of each grating is reduced in comparison with the object grating. The contrast is the ratio of the local maximum to minimum intensity (I_{max}/I_{min}). The MTF is the ratio of the difference between I_{max} and I_{min} in the image plane, normalised with respect to the difference between I_{max} and I_{min} in the object plane. The orientation of the lines for lens testing is usually perpendicular to two orthogonal axes, as in the 1951 U.S. Air Force (USAF) test chart. The resolution of the lens is described by its 50% MTF which is the spatial frequency at which the MTF is 50% of its low frequency value. Companies such as Image Science Ltd offer equipment for measuring the MTF of lenses.

Point source inputs are an alternative to Ronchi gratings. The results of point source testing are of interest in the case of lenses for astronomical telescopes because distant stars behave as point sources. In this case, the squared magnitude of the PSF is used for the assessment of lens quality. The diffraction pattern of a circular aperture is an Airy disc or pattern [2]. If the system is not diffraction limited, then the diffraction pattern includes the additional effect of aberrations in the lens. In this case, the diffraction pattern spreads, and the peak amplitude decreases. The decrease in the intensity of the peak has been put on a quantitative basis in the Strehl ratio (SR). The SR is the peak intensity of the actual image of the point source, normalised with respect to the diffraction limited image of the point source. It is found that the SR correlates well with the normalised power of the light in the central lobe of the aberrated image of the point source. A familiar example of the use of point sources is perimetry testing of the visual field.

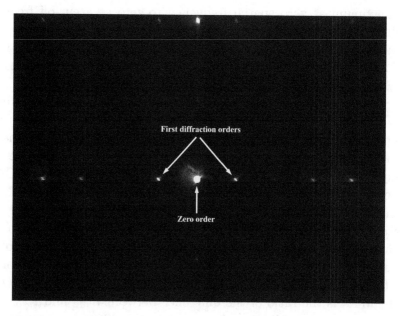

Figure 4.1 The diffraction from a vertical Ronchi grating of four pixel period on an EASLM

Transfer functions can also be defined and measured in the case of EASLMs and image sensors. The modulation transfer in these cases is not between object and image planes. It is between an electronic representation of the object and its optical representation in the first case, and vice versa in the second case. This entails that the MTF is quantified in a relative sense, and normalised with respect to low spatial frequency. Relevant objects are Ronchi gratings for EASLM testing and lines for image sensor testing. EASLMs have a larger pixel repeat than image sensors. Therefore, the MTF of an amplitude modulating EASLM can be measured directly by a good quality camera employing a high quality lens. Phase modulating EASLMs are measured on the basis of diffraction efficiency. The diffraction efficiency of a grating is the power of light diffracted into the first diffraction order normalised by the power of light in the incident beam. The diffraction from a Ronchi grating of four pixel period on an EASLM illustrates the first order diffraction (Figure 4.1). The most significant spot is the zero order corresponding to light which has not been diffracted. The diffracted spot in the centre of the top of the image is due to the inherent pixellation of the EASLM. There are corresponding spots on the left and right extremes of the image. The first diffraction orders lie on either side of the zero order. They are situated at one quarter of the distance to the diffracted spots due to the pixellation, since there are four pixels per period of the grating. They are known as the +1 and -1 orders. The second order diffraction spots are missing since the grating is symmetric. The third orders show faintly because their intensity is 1/9th of the first order intensity. Finally, the intensity of the pixellated diffraction in the horizontal direction is reduced because it coincides with an even harmonic of the grating period, which is suppressed due to the symmetry of the grating. In order to measure the line response function of an image sensor, a rectangular slit is imaged onto the device. Due to the small pixel repeat of image sensors, the diffraction spreading of the slit imaging system is included in the response function. This is removed by deconvolution (Section 1.3.3).

When an image is sampled by an EASLM or an image sensor, there is an upper limit to the spatial frequency representation, which is $(2\Delta x)^{-1}$, where Δx is the pixel repeat in the x-direction. If the spatial frequency content of the image exceeds this, then aliasing results. Aliasing is the confusion between the actual pattern and a similar pattern of lower resolution. In the time domain, this is the familiar example of sampling a spinning wheel with a lower frequency strobe light. The wheel appears to move slowly and may be observed rotating in the opposite sense. In Fourier optics, a comparable example is the use of an SLM to represent a lens function in an optical system. For short focal length lenses where the optical path changes rapidly at the periphery of the lens, the pixel repeat of the SLM can be insufficient. The effect of this is that the SLM samples the lens structure and this has the effect of replicating the lens in the image plane of the encoded lens [40]. In order to represent a given image with no aliasing, the cut-off frequency of the device should be at least twice the highest spatial frequency of the image. In order to define the highest spatial frequency of an image, it is digital picture functions that are considered, as captured on an image sensor with small pixel spacing. This is then known as a bandlimited image, although the band limitation arises from the image sensor.

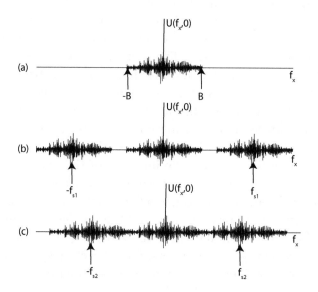

Figure 4.2 The real part of the x-component of the spatial frequency spectrum of an image:(a) unsampled; (b) sampled at spatial frequency f_{s1}; (c) sampled at spatial frequency f_{s2}

Sampling

The rule-of-thumb that the sampling frequency of an image should be at least twice the highest spatial frequency of the image is sometimes referred to as the Nyquist criterion [84] or the Whittaker-Shannon theorem [90]. A 1D spatial frequency spectrum of the real values of the Fourier transform of an image is shown in Figure 4.2(a). When this spectrum is sampled, this results in the appearance of higher orders, which are mentioned in Section 2.1.1 and illustrated as the diffracted spots due to the pixellation in Figure 4.1. When the sampling frequency, f_{s1}, is greater than 2B, where B is the maximum frequency of the spectrum of Figure 4.2(a), the higher orders are separate from the original spatial frequency spectrum (Figure 4.2(b)). The original object can be reconstructed accurately by low pass filtering this replicated spectrum. However, when the sampling frequency, f_{s2}, is less than 2B, the higher orders are not clearly separated from the original spatial frequency spectrum (Figure 4.2(c)). In the border region between the original spectrum and the higher order, it is uncertain whether the frequency value corresponds to a frequency in the original spectrum or in the higher order. This is the aliasing due to an insufficient sampling frequency.

4.3   IMAGE SENSOR

Image sensors have been developed for visible and infrared wavelengths from 0.3 up to 16.5 μm. The infrared sensors are known as focal plane arrays, and will not be covered here. The visible region sensors considered here are based on single-crystal silicon. The silicon can be doped with an electron acceptor, such as boron, to create a material known as p-type silicon. Then the surface of the crystal is thermally oxidized to give a thin layer of SiO_2 and metal electrodes are evaporated on top of the SiO_2. A positive voltage on the metal then creates a layer in the silicon which is depleted of charges. Upon illuminating this layer, a pocket of electrons is formed whose number is proportional to the intensity of the radiation. This pocket of electrons can be moved across the surface of the silicon by clocking the positive voltage on neighbouring electrodes. This is the basis for the charge-coupled device array (CCD), which was invented in 1969, and its impact on society was such that the inventors, Willard Boyle and George E. Smith, received the Nobel Prize in 2009. Their idea was to use the CCD as a computer memory chip. Their colleague, Mike Tompsett, saw its potential as an image sensor. Sony introduced the CCD camera in the late 1980s and stopped production in 2016. The photocharge accumulated at each pixel of the CCD is clocked across the sensor to an output amplifier where it is converted to an analogue voltage initially, followed by analogue-to-digital conversion (ADC). A high quality image results with low fixed pattern noise and low dark current. CCDs achieve high sensitivity as well as good signal quality in low light conditions due to their high fill factor.

The method for clocking the charge across the CCD can be full frame, frame transfer, or interline transfer. The KAF-1603 and the

FTF7046M are full frame devices and the KAI-29050 is interline transfer (Table 4.1). The full frame CCD clocks the photocharge line by line to a shift register. Consequently, there is charge smearing during the transfer from the light-sensitive to the masked regions of the CCD. It is preferable to use a light shutter with the full frame device. These devices offer 100% Fill Factor but use less silicon so that they are less expensive. The frame transfer CCD has the sensitivity of the full frame device but it is typically more expensive due to the larger sensor size needed to accommodate the frame storage region. Although the frame storage region is protected from incident light, there is still smearing during the readout time. However, the readout time is shorter than for the full frame CCD, because the full sensor is readout at once rather than line by line. Interline devices have a charge storage area next to every pixel. This effectively reduces the light sensitive area of the sensor, so that the Fill Factor is reduced. This can be partially compensated by the use of microlens arrays to focus the received light on the photosensor area. The compensation usually works best for parallel light illumination but for some applications which need wide angle illumination the sensitivity is significantly compromised. The interline-transfer CCD incorporates charge transfer channels called Interline Masks. These are immediately adjacent to each photodiode so that the accumulated charge can be rapidly shifted into the channels after image acquisition has been completed. The very rapid image acquisition virtually eliminates image smear.

Despite its proud history, several disadvantages of the CCD have been appreciated. In order to display images on analogue monitors, the early CCDs were equipped with interlaced scanning. Progressive scanning is the technique of choice for coupling into EASLMs and PCs. Frame transfer and full frame CCDs are inherently progressive scan, but interline transfer CCDs can be either progressive or interlaced. A further disadvantage is the limited readout speed, due to the serial output data stream. In Table 6.1, the speeds of the CCD sensors are quoted in terms of the serial output data rate. Further disadvantages of CCD sensors are blooming and smearing. Blooming occurs when the photogenerated charges exceed the capacity of the pixel and spill over into adjacent pixels. Smearing is the generation of additional charges during the readout process, which is an issue in the full frame devices. The mainstream image sensor nowadays is the CMOS array.

The invention of MOS image sensors preceded that of the CCD [185], but the lower fixed-pattern noise of CCDs gained the day. Advances in CMOS processing and further work by researchers at the University of Edinburgh and Linkoping University led to this image sensor gaining ground [81]. The first company to market the product, VLSI Vision, was sold to ST Microelectronics in 1998. In the CMOS sensor, the charge to voltage conversion is performed within each pixel. This requires more complex electronics at the pixel, e.g., 4 or 5 transistors per pixel, and non-uniformity of the circuitry between the pixels leads to fixed pattern noise. Parallel readout of the image information from a CMOS sensor offers the advantage of higher frame rates at comparable resolutions to the CCD. Direct addressing of individual pixels allows the definition of regions of interest. The higher level of integration makes the CMOS camera more cost effective with reduced power consumption in comparison to CCDs. A further advantage is the concurrent sampling of the signal on all pixels (called global shuttering). A minor disadvantage is that charge to voltage conversion at the pixel reduces the pixel real estate available for charge collection. Therefore, the CMOS sensor has a reduced dynamic range compared with the CCD.

The dynamic range (DR) of the sensor is the contrast between the brightest and the weakest measurable light signal in the same image. The DR is defined in the European Machine Vision Association (EMVA) 1288 standard for characterisation of image sensors and cameras [73]. It is equal to the ratio of the signal saturation, $\mu_{p,sat}$, to the absolute sensitivity threshold, $\mu_{p,min}$, where the subscript p refers to photons. The sensor manufacturers provide the data for the full well capacity and the read noise (Table 4.1). The full well capacity sets the upper limit for the brightest signal, and the read noise sets the lower detection limit. These values correspond, approximately, to $\mu_{e,sat}$ and $\mu_{e,min}$ in the EMVA standard, respectively. The standard distinguishes between full well capacity and the saturation capacity represented by $\mu_{e,sat}$. The latter is slightly less than the full well capacity. Since $\mu_{e,sat} = \eta\mu_{p,sat}$, where η is the quantum efficiency, and $\mu_{e,min} = \eta\mu_{p,min}$, the DR is approximately equal to the ratio of full well capacity to readout noise. It is usually quoted in dB units, which is 20 times the logarithm of this ratio.

Related to the DR is the bit depth of the voltage ascribed to each pixel. A useful analogy is that the DR is the height of a staircase and the bit depth is the number of steps. Most sensors use ADCs with a resolution between 8- and 14-bits. Some devices have a fixed resolution. For example, the resolution of the LUPA1300 in Table 4.1 is 10-bit resolution, whereas the CMV50000 is 12-bit resolution. Although a higher bit depth would seem preferable, this impacts the frame rate. The IMX253 runs at a frame rate of 46.4 fps at a bit depth of 12-bit. However, it runs at 68.3 fps at a bit depth of 8-bit.

A good measure of the sensitivity of the sensor is the signal to noise ratio (SNR). The ideal is to operate a photosensor with sufficient illumination so that it is photoelectron shot noise limited. This shot noise is proportional to the square root of the number of photogenerated electrons. Since the signal is proportional to the number of photogenerated electrons, the ratio of signal to noise, or SNR, is proportional to the square root of the number of photogenerated electrons. Therefore, the maximum sensitivity is the root of the maximum electron capacity of the pixel, $\mu_{e,sat}$. When the sensor is used for image capture in low light applications, the noise floor is determined by electrical noise. In the scientific CMOS (sCMOS) sensors, a unique architecture of dual-level amplifiers and dual ADCs helps maximize dynamic range and minimize readout noise at the same time. For example, the CIS1910A has a full well capacity of around 30,000 electrons and a 1.2 electron rms readout noise in rolling shutter mode and 4 electron rms readout noise in global shutter mode (see later for the difference between the shutter modes). In the former case, the DR is 88 dB and, in the latter case, it is 78 dB. The noise at a single pixel can be reduced by decreasing the temperature of the sensor. The DS-Qi2 employs electronic cooling for this purpose.

It is also important to reduce the variations between individual pixels and deliver a homogeneous and stable dark image. The dark signal non-uniformity (DSNU) value, which is the standard deviation of the dark noise over all the pixels of the array, can be well below $1e$ rms in the best sensors. Another source of noise is the photon response non-uniformity (PRNU). This is the root mean square (rms) difference between the individual 50% saturation signals and the dark signals, normalised by the difference between the mean 50% signal and the dark signal, as defined in EMVA 1288 standard. PRNU values measured for sCMOS cameras are generally below 0.5%, indicating that neighbouring pixels differ by less than 0.5% in their response to the same light signal.

In general, CCD-based cameras offer high sensitivity but slow sampling speeds. Conventional CMOS cameras offer very fast frame rates but compromise dynamic range. sCMOS image sensors, on the other hand, offer extremely low noise, rapid frame rates, wide dynamic range,

TABLE 4.1 Exemplar image sensors and one compact camera

Manufacturer	Product	Type	Array Size	Resolution	Pixel pitch	Read noise	Full well capacity	Frame/ Data rate
CMOSIS	CMV50000	CMOS	36.4 x 27.6 mm	7920x6004	4.6	8.5e	14,000e	30 fps
CMOSIS	Nanoeye GS	CMOS	3.4 x 3.4 mm	640x640	3.6	16e	16,000e	100 fps
Dalsa	FTF7046M	CCD	36 x 24 mm	6936x4616	5.2	11e	40,000e	1.2 fps
e2v	EV75C776	CMOS	11.5 x 11.5 mm	4096x4096	2.8	2e	7,000e	45 fps
Fairchild	CIS1910A	sCMOS2.0	12.5 x 7 mm	1920x1080	6.5	4e	30,000e	50 fps
Fairchild	LTN4625A	sCMOS2.0	25.3 x 14.3 mm	4608x2592	5.5	5e	40,000e	120 fps
Nikon	DS-Qi2	CMOS	36 x 23.9 mm	4908x3264	7.3	2.2e	60,000e	6 fps
ON semi.	KAF-1603	CCD	13.8 x 9.2 mm	1536x1024	9	15e	100,000e	10 MHz
ON semi.	KAI-29050	CCD	36.2 x 24.1 mm	6644x4408	5.5	12e	20,000e	40 MHz
ON semi.	LUPA1300-2	CMOS	1.4″ diag	1280x1024	14	37e	30,000e	500 fps
Sony	IMX253	CMOS	17.6 mm diag	4112x3008	3.45	3e	10,000e	46.4 fps

high quantum efficiency, high resolution, and a large field of view simultaneously in one image. This makes them particularly suitable for high fidelity, quantitative scientific measurement in low-light-level conditions. With regard to potential applications in Fourier optical systems, the high sensitivity of the CCD camera would make it appropriate for scene capture. The flexibility of the CMOS camera is important for processing within and at the output of the system, where, in addition, the provision of an anti-reflection (AR) coating, which prevents reflection within the system, is important. The application can also be influenced by the readout mode for the camera: either rolling or global shuttering. The rolling shutter allows the readout of columns of pixels to be staggered, which offers lower noise and higher frame rate than global shutter. However, this restricts the image capture application to static or slow moving objects. Therefore, the performance data for CMOS sensors in Table 4.1 refers to global shuttering, where the whole frame is captured at the same time.

An alternative to the CCD and CMOS sensors is the position sensitive detector (PSD) and the profile sensor (PS). The PSD locates the beam position on a 2D photodiode surface by locating the electrodes at the periphery of the surface [250]. An example is the S1880 sensor from Hamamatsu. The PS is a photodiode array which is specialised for peak detection [174]. Each pixel is divided into two sub-pixels, called X and Y. All the X-pixels in a row are connected together, and all the Y-pixels in a column are connected together. A scan of the rows gives the X location of the peak intensity, and a scan of the columns gives the Y location of the peak intensity. The speed of the PS sensor is high, 800 fps for the 512x512 pixel version based on a 10 bit ADC.

4.4 CAMERAS

Modern cameras deliver a very homogeneous and stable dark image which is beneficial especially in low light applications. The speed of simple camera processing operations has been improved by the use of FPGA based frame grabber cards. The low-voltage differential signalling (LVDS) interface, which is supported by many image sensor manufacturers in their products, specifies the electrical characteristics of a differential, serial communications protocol, to enable camera developers to route the data output from a CMOS image sensor directly into an FPGA. Each pair of such signals enables data to be transferred from the image sensor to the FPGA at rates at about 600 Mbit/sec.

Separate clock sources are used to enable the FPGA to accurately recover synchronized data from the imager. LVDS channels have a low susceptibility to noise because sources of noise add the same amount of common-mode voltage to both lines in a signal pair. The use of LVDS interface on sensors, in general, reduces power consumption, which is important when the power budget is critical.

Sony have developed a new standard called Scalable Low Voltage Signalling with an Embedded Clock (SLVS-EC) interface, which allows a high-speed interface with a lower pin count. As a result, the cost of building cameras with high resolution and high data rates will be lower than with the LVDS interface. The SLVS technique is based on a point-to-point signalling method that has evolved from the traditional LVDS standard. SLVS also has a low susceptibility to noise, but because the specification calls for smaller voltage swings and a lower common-mode voltage, lower power drive circuitry can be used to drive the interface. In the SLVS-EC design, the clock signal is embedded in the data from the imager and recovered by dedicated circuitry on the FPGA. Hence the data can be transmitted at much higher data rates and over much further distances. Specifically, each of the channels on the SLVS-EC interface can support effective data rates of over 2 Gbit/sec.

In order to realise low-cost optical systems, it will be important to fabricate cameras at low cost. Wafer scale camera technology is aimed at fabricating the image sensor and associated lens in volume by processing 8 inch diameter silicon wafers and multilayer wafer-scale lens assemblies in mask aligner systems. The complete cameras can be subsequently diced and separated. The Omnivision OVM7690 is a single chip image sensor, embedded processor and wafer-level optics in one compact, small-profile package of 2.5 x 2.9 x 2.5 mm. The CMOS image sensor (CMOSIS) NanEye module has a camera volume of only 1.5 mm^3 and a low power dissipation, which is important for remote systems. Details of the larger NanEye GS are included in Table 4.1. For most measurement applications, it is very convenient to have a linear response of the output signal to changes in light input. Deviations from such a linear behavior are addressed as non-linearity. Cameras employ the necessary electronic components and firmware algorithms to correct the non-linear behavior in real time providing for linearity better than 99%.

GLOSSARY

Airy pattern or disc: This is the diffraction pattern of a circular aperture, consisting of a central lobe surrounded by a sequence of rings of lower intensity.

Aliasing: The ambiguity caused by insufficient resolution on an EASLM or an image sensor to display the required pattern.

Dark signal non-uniformity: DSNU is defined as the spatial signal variation between pixels without illumination.

Deconvolution: The converse of convolution in that the output is divided by the spread function in order to derive the input function.

Photon shot noise: The number of photons collected by a camera pixel over a given time interval is subject to random statistical fluctuations. The variability of the number is the photon shot noise.

Read noise: Noise associated with all circuitry that measures and converts the voltage on a pixel into an output signal.

Light sources

CONTENTS

5.1 INTRODUCTION

The principal Fourier optical system architectures have been demonstrated using high coherence length gas lasers. In the interest of building compact, low-cost systems, the sources of choice are the semiconductor diode laser (DL) and light emitting diodes (LEDs). The DL was invented in 1962 by Robert Hall of General Electric Corporation. His team assembled a device from pieces of p- and n-type gallium arsenide. When these are polished and contacted, and one face of the sandwich is polished, the p-n junction is ready for current injection. If the p-type material is the anode and the n-type the cathode, then current can be injected into the p-n junction diode with low resistance. Coherent laser radiation is emitted from the polished face of the p-n junction when the injected current exceeds a threshold level. Hall's device required a 10 amp current at liquid nitrogen temperatures. The current was delivered in a 1 to 10 microsecond pulse, in order to avoid over-heating the sandwich. Important advances in both the material and architecture of the laser have been made in the last 50 years. The threshold current has been reduced to sub-milliamp levels in laboratory devices, although tens of milliamps is more customary for commercial devices. In order to access applications spanning blu-ray disc readers and automobile headlamps to fibre optic communication, a wavelength range from 405 to 1550 nm has been accessed by material selection. The global laser market was worth over \$8 billion in 2013, of which 50% was the diode laser market. The focus here will be on the red region between 630 nm and 700 nm and near infrared region between 705 nm and 850 nm. These lasers are now well established, with good power and lifetime characteristics, and match well with the spectral response of image sensors based on silicon photodiodes.

Red diode lasers are fabricated by growing layers of ternary and quaternary compounds of aluminium, gallium, indium, and phosphorus on GaAs wafers. The layers are sandwiched into a p-n junction which is formed by ion implantation on the upper and lower layers. Subsequent to the processing they are cleaved and diced into individual lasers of typically 500 μm length, 100 μm width, and several microns thickness. The junction runs along the width of the laser. The cleaving produces good quality facets at the two ends of the length, so that an optical cavity is formed where the radiation is reflected at the two ends and confined to a narrow strip. This basic laser structure is called an edge emitting laser, because the radiant emission is from the cleaved face at the narrow edge of the laser die. The forward current in the diode stimulates the emission of radiation with an energy just above the bandgap energy of the semiconductor material. The stimulated emission exits from the front facet of which the reflectivity is around 30 to 40%.

The interest here is in single transverse mode lasers, also known as TEM_{00}, where the aperture of the laser is so small that higher order modes are not sustained. This mode has a constant phase wavefront, of which the intensity profile, $I(x, y, z)$, is Gaussian

$$I(x, y, z) = I(0, 0, z) exp\left[-\left(\frac{2x^2}{\omega_x(z)} + \frac{2y^2}{\omega_y(z)}\right)\right], \qquad (5.1)$$

where $\omega_x(z)$, $\omega_y(z)$ are the Gaussian beam radii in the x- and y-directions at a distance z from the beam waist. The beam waist of the diode laser, where the Gaussian beam radii take the values ω_{x0} and ω_{y0}, is located at the diode. The Gaussian beam radius is the radius at which the intensity of the beam has decreased to $\frac{1}{e^2}$ of its on-axis value. Due to the small aperture of emission, the radiation is emitted in a diverging beam of, typically, 7 degrees parallel to the junction, and 12 degrees perpendicular to the junction. The divergence angle here is the angle subtended between the axis of the beam and the Gaussian beam radius. The divergence angles can be expressed in terms of the Gaussian beam radii at the waist

$$\theta_\| = \frac{\lambda}{\pi \omega_{x0}}, \qquad \theta_\perp = \frac{\lambda}{\pi \omega_{y0}}. \tag{5.2}$$

Therefore, the brightness of a laser of power, P, and wavelength, λ, with a diffraction limited output beam is given by

$$B = \frac{P}{\pi \omega_{x0} \omega_{y0} \pi \theta_\| \theta_\perp} = \frac{P}{\lambda^2}. \tag{5.3}$$

where (5.2) has been used. The units for the brightness are $W sr^{-1} m^{-2}$.

The centre wavelength of the laser is an important parameter because this scales the size of the FT. There is a $\pm 0.5\%$ to 1.5% variability of the centre wavelength of high power laser diodes, due to fabrication tolerances. If the temperature of the laser mounting is allowed to increase to around 60 deg. C, this can result in an additional $+1\%$ increase in the centre wavelength of the emission. Moreover, the light emission is composed of a range of wavelengths around the centre wavelength, and this is the spectral width of the emission, or linewidth. This is quantified by the full width half maximum (FWHM), which is the width of the spectral distribution at the intensity which is one half the maximum intensity. The FWHM of diode lasers is, typically, a few nanometres.

The LED was the pre-cursor of the DL. The story of the LED extends back to 1907, when Round discovered electroluminescence [292]. The first commercial red LED was made at Ferranti's Photon Devices Group [202]. Improvements in the LED have led to applications in domestic and automobile lighting. Although the brightness of an LED is orders of magnitude less than that of the DL and the linewidth is about an order of magnitude larger, this disadvantage is offset by the much lower price and the longer lifetime. Future application in Fourier optical systems will require careful design in order to comply with the coherence requirements of the system.

5.2 COHERENCE

In order to describe light propagation using wavefronts, the vibrations of the electric field at two neighbouring points in space and time must be correlated. Coherence is a measure of this correlation in both the spatial and temporal dimensions. A fully coherent beam is one where correlation extends over space and time, for example a laser beam. A wavefront links points in space where the electric field vectors vibrate in phase at an instant of time. The spatial coherence of the beam is a measure of the lateral extent of the wavefront. At a given point in space, the wavefronts pass regularly until there is a disruption of the phase. The distance between one disruption and the next, averaged over a sufficient number of points, is the coherence length of the laser which is a measure of the temporal coherence of the beam. Lasers are known for being sources of high brightness and coherence. The light field of a laser can be described by a series of wavefronts which extend without disruption over the coherence length of the laser in the direction of propagation. The coherence length is inversely proportional to the linewidth of the laser. On dimensional grounds, the coherence length is given by $\lambda^2 \Delta\lambda^{-1}$, where $\Delta\lambda$ is the FWHM of the laser. More exact calculation gives a pre-factor which depends on the exact spectral profile of the linewidth. For example, this may lie between 0.44 and 0.66. For a 730 nm diode laser with a linewidth of 2 nm, the coherence length is between 120 and 170 microns. It can usually be calculated from the manufacturer's specification of the laser. The spatial coherence is a parameter which must be measured by experiment.

The effects of coherence had been studied for a century before the discovery of the laser. The effects of partial coherence of extended light sources were studied in different contexts [26]. The subject of coherence was considerably simplified when Zernike defined a quantitative measure, namely the degree of coherence, by a measurement procedure, namely the visibility of fringes in a double-slit experiment (Section 5.2.1). The visibility of the fringes is the difference between I_{max} and I_{min}, normalised by the sum of I_{max} and I_{min}. Therefore, the numerical value of the degree of coherence is bounded by 0 and 1, It was shown by van Cittert that this definition is simply the ratio of intensity of fully coherent light to the total light intensity [263]. The total light intensity in this model is the sum of coherent and incoherent light intensities. In the same paper, van Cittert reiterated his earlier explanation of the cause for the partial coherence of extended sources. The explanation follows from a consideration of the overlap of the probability distributions of the amplitude and phase of the light field at the two locations considered. The mathematical complexity of the explanation did not allow further experimental investigation, compared with the simpler treatment of Zernike. Therefore, the Zernike treatment is used here to characterise the degree of coherence of the laser. A degree of coherence close to one equates to a laser beam where the surfaces of constant phase (wavefronts) are smooth and extend over the size of the devices and elements used in the optical system. The degree of coherence of a diode laser source depends on both the size of the diode and the quality of the collimation optics.

In order to represent the correlation between the amplitude and phase of the light field at two locations in space, the mutual coherence function, $\Gamma_{mn}(\tau)$ is used, where

$$\Gamma_{mn}(\tau) = \langle U_m(t)U_n^*(t+\tau)\rangle, \tag{5.4}$$

$U_m(t)$ is complex electric field amplitude at position r_m and time t. The $\langle\ \rangle$ brackets indicate that the product of the field amplitudes is integrated over the time period of the waveform, and divided by the value of that time period.

Then, the complex degree of coherence is given by

$$\gamma_{mn}(\tau) = \frac{\Gamma_{mn}(\tau)}{\sqrt{\Gamma_{mm}(0)}\sqrt{\Gamma_{nn}(0)}}, \tag{5.5}$$

and the absolute value of $\gamma_{mn}(0)$ is the value which was measured in the fringe visibility experiment.

Finally, the mutual intensity, J_{mn} is equal to $\Gamma_{mn}(0)$.

5.2.1 Double-slit experiment

The double-slit experiment was performed originally by Thomas Young in 1801 using sunlight, which has a significant degree of coherence because of the small angle which it subtends at the earth's surface. The experiment is commonly called Young's slits. The slits sample the beam at two locations along one axis. In order to measure the coherence along the orthogonal axis, it is important to re-orient the slits. In view of the need to change the spacing and orientation of the slits, it is convenient to use an amplitude modulating EASLM to define the slits. The slits define secondary sources for the beam. For two slits separated by a distance, D, along the x-axis, the total complex amplitude after the slits is the sum of the aperture functions of each slit, assuming that the slits are illuminated equally

$$U(x,0) = rect(x - D/2) + rect(x + D/2). \tag{5.6}$$

The diffraction pattern along the x'-direction at a distance z_0 is given by $U(x', z_0)$ where

$$U(x', z_0) = \frac{e^{ikz_0}}{i\lambda z_0} \int U(x,0) exp\left[ik\left(-\frac{x'x}{z_0}\right)\right] dx. \qquad (5.7)$$

Using the Fourier shift theorem (Section 1.3.4)

$$U(x', z_0) = \frac{e^{ikz_0}}{i\lambda z_0} \left(sinc(f_x)e^{-i\pi f_x D} + sinc(f_x)e^{i\pi f_x D} \right) \qquad (5.8)$$

$$U(x', z_0) = \frac{e^{ikz_0}}{i\lambda z_0} 2sinc(f_x)cos(\pi f_x D), \qquad (5.9)$$

where $x' = f_x \lambda z_0$.

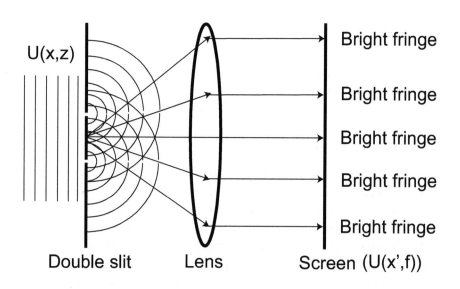

Figure 5.1 Double-slit experiment: wavefronts and raytrace

This produces an interference fringe pattern of cosine squared fringes in the focal plane of a lens placed after the slits. The arrival of wavefronts at the slits and the creation of wavelets by the slits is illustrated in Figure 5.1. Five rays are traced from the plane of the slits to the output screen. These correspond to the five directions where the wavelets sum to produce constructive interference and bright fringes are formed on the output screen. Additional directions at larger angles which also produce fringes have not been traced, in order to preserve clarity. The intensity of the fringes is modulated by a $sinc^2$ function, so that the intensity of these additional fringes is low. Figure 5.2 shows the results of a double-slit experiment using a cooled solid state red laser diode with $D = 1.3$ mm. When the source plus collimation optics has a degree of coherence less than one, then lower visibility will be observed in the fringe pattern. This will be manifested by an increase in the value of I_{min}, i.e., the dark state will be less dark. In order that a source be suitable for a Fourier optical system, there should be no decrease in visibility of the fringes for slit spacings, D, up to the aperture of the optical devices which will be used. At the spacing of the slits shown in Figure 5.2a, the fringe visibility is maintained close to 1. The minima of the fringes is close to that expected as a result of dark noise. One of the requirements for a good measurement with this set-up is that stray light be reduced to a minimum. It is expected that a larger spacing of the slits will result in a reduced fringe visibility. The separation of the slits at the position of reduced visibility will define the maximum aperture of the EASLM which can be used with the source plus collimation optics. The maximum aperture for current EASLMs is $1.55''$ diagonal (see Table 2.1).

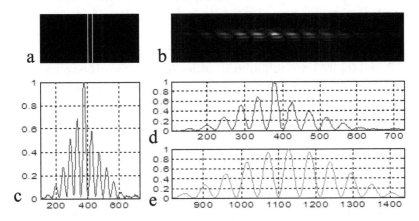

Figure 5.2 Double-slit experiment: (a) image of slits; (b) image of diffraction pattern; (c) cross-section of (b); (d) magnified cross-section; (e) simulated fringe pattern

5.3 LASER DIODE OPTICS

Due to the relatively large angle emission, the optics required to collimate the laser diode must have a low f-number for high efficiency. Moreover, the wavefront distortion of the collimated beam must be low so that the wavefront is effectively flat. Fortunately, a good laser diode distributor will sell collimator assemblies for the stock lasers. In addition to collimation, it is desirable to transform the elliptical cross-section of the beam to a spherical one and to re-distribute the energy in the beam so that it is uniform. The circularisation of the beam is conventionally performed by an anamorphic prism pair which is also sold by some distributors. The re-distribution of energy can be performed by a DOE (Section 3.2), or by using refractive laser beam reshapers [103].

5.4 VERTICAL CAVITY SURFACE EMITTING LASER

Practical vertical cavity surface emitting laser (VCSEL) structures were invented in 1989 [124]. In contrast to the edge emitting laser, the VCSELs emit perpendicular to the face of the wafer. This increases the complexity of the fabrication technology because, in addition to the ion implantation for the electrical contacts, high reflectivity mirrors must be grown on both the front and back faces of the wafer. The cavity length is three orders of magnitude less than that of the edge emitting laser. This reduces the stimulated emission gain of the beam when it traverses the cavity. Therefore, higher reflectivity mirrors (between 99.5 and 99.9%) are required in order to maintain the round trip gain. The short cavity length does improve the temporal coherence, and the single transverse mode VCSELs have a single longitudinal mode, and, therefore, a long coherence length. The beam output is circular but the power is generally lower than the edge emitter due to the lower round trip gain. The main advantages of VCSELs are that the operation is relatively temperature insensitive, the threshold current is low, and the architecture is amenable to wafer-scale production. Although they are not used in Fourier optical systems at present, wafer-scale production is promising for future, low-cost optical processing microsystems.

5.5   LIGHT EMITTING DIODE

Contemporary light emitting diodes (LEDs) range from high-power, multi-die sources to low-power, point source emitters. Of the high-power sources, a large number have to be rejected for this application because they consist of blue LED emitters with an encapsulation containing a phosphor which supplements the blue wavelength with yellow to provide a cold white light. An example of a small form factor, high-power, monochromatic source is the Lumileds Luxeon Z Color Line emitter. Its size is 1.3 mm x 1.7 mm and the linewidth is 20 nm. An example of a point emitter is the Marktech MTPS7065MT with a linewidth of 15 nm and a beam aperture of 150 μm. The power output of the Marktech is two orders of magnitude less than the Luxeon source. The photometric intensity of these LEDs in a given direction approximates Lambert's cosine law [26]. The photometric intensity, in units of Wm^{-2}, is the integral of the brightness of the source over the solid angle. The spatial coherence of the Luxeon source can be estimated from application of the van Cittert-Zernike theorem [26]. This theorem states that the mutual intensity of the emitted wavefront is equal to the Fourier transform of the aperture of the source, provided that the LED is spatially incoherent across its aperture. If it is assumed that the latter is the case, the FT gives a *sinc* function with a zero when $kpa = \pi$, where k is the magnitude of the wave vector, p is the direction cosine of the diffracted ray, and $2a$ is the linear dimension of the LED aperture. This equation can be solved for p with the result that $p = \lambda/2a$. The propagating beam from the LED will have a coherence area perpendicular to the beam at a distance, z, with a size given by pz. It can be seen that, for a wavelength of λ=0.633 μm, that the beam must be propagated for 1 metre to give a coherence area of 0.5 x 0.4 mm^2. This is clearly inefficient, especially since most of the light flux from the LED will be outside this area. In the case of the Marktech LED, the FT of the circular aperture gives an Airy pattern. The distance between the first zeroes in the Airy pattern is $2.44z\lambda/2a$, which gives a coherence area equal to a circle of diameter 10 mm for a propagation distance of 1 metre. This is an improvement, but at the cost of reduced power. The Lambertian nature of the emission necessitates high f-number lenses in order to capture a good fraction of the emitted power. A number of designs have been made in this area, such as the total internal reflection (TIR) lens [198], the lens with a light collecting cone mirror [285], surface shaping using orthogonal polynomials [28],

3D reflectors [91, 123, 216], aplanatic designs [283], and many others. The main emphasis has been on maximising flux transfer rather than the coherence of the light output. Superluminescent LEDs (SLEDs) have a small source size, but the power is low and the cost is high, so they are currently not suitable for the systems discussed here.

GLOSSARY

Coherence length: The distance along a ray over which a defined phase relationship between the complex field amplitudes is maintained.

Spatial coherence: The property of a light field covering a region of space where the phase relationship between the field at different points is well defined at one instant of time.

Temporal coherence: The property of a light field along a ray where the phase relationship between the field at different positions on the ray is well defined.

CHAPTER 6

Optical correlators

CONTENTS

6.1 INTRODUCTION

The primary application of Fourier optics in image processing is the optical correlator system. The system design involves the calculations of both ray optics and diffraction integrals. This is the subject of Section 6.2. In order to construct a correlator system, an essential element was the complex spatial filter. Vander Lugt developed a technique for recording a complex spatial filter in 1966 and constructed the first correlator based on this in the same year [264]. The spatial filter in this correlator was recorded on a holographic recording material, which had to be processed and replaced in the system after processing. Holographic recording materials are discussed in Section 6.3. The Vander Lugt correlator (VLC) architecture and its variants are discussed in Section 6.4. In order to avoid the high precision required in the alignment of the spatial filter in the VLC, a novel correlator architecture was proposed and demonstrated [280]. This became known as the joint

transform correlator (JTC) (Section 6.5). The third correlator architecture, which has been extensively explored since then, is the hybrid
correlator, where significant electronic processing is involved (Section
6.6).

6.2   SYSTEM ANALYSIS

In order to design the Fourier lens both with regard to the exact focal
length and the reduction of aberrations, geometric ray tracing must be
adopted. The placement of the Fourier lens with respect to the SLMs
and other optical components influences the Fourier transform. Since
the transform involves a diffraction integral, it is important to be able
to manipulate these integrals efficiently. The mathematics involved in
both these aspects will be discussed here.

6.2.1   Geometric optics

If the Fourier optical system can be designed with no constraints on the
choice of Fourier lens, then achromatic doublets are preferred. These
are computer designed to effectively minimize spherical aberration and
coma when operating at an infinite conjugate ratio, i.e. for an object
placed in the focal plane of the lens. The aperture of the lens should
be double the diagonal aperture of the SLM. The lens is composed of a
positive and negative lens, oriented so that the positive lens faces the
object plane. An example is the doublet sold by Space Optics Research
Labs, LLC for their FX 15/5F Fourier system, which is an f/5 lens
with a 7.6 cm diameter and 38 cm focal length. Over the central 3.8
cm aperture, the resolution is 32 lp/mm and the wavefront accuracy is
$\lambda/8$. The effective SBWP of this lens in the central aperture exceeds
10^6.

If the selection of the Fourier lens is constrained, so that the focal
length must be achieved by a combination of lenses, then the focal
length of the combination must be computed. A convenient method
for following the path of a light beam through a sequence of lenses is
the use of ray matrices. The ray is defined in two dimensions, the optic
axis, z, and one axis perpendicular to this, x, say. At a particular value
of z, the ray will be defined by two coordinates, x and θ, where θ is
the angle of the ray direction with respect to the optical axis, z. In
order to follow the progression of the ray through the lenses, these two
parameters are collected in a column vector, $[x \quad \theta]$. The propagation

of the ray through a distance, d, results in the augmentation of x by $d\theta$, whereas θ remains unchanged. The use of the angle, θ, rather than $sin\theta$, is permitted provided that the angles are small enough to make this approximation. This is called first order or paraxial optics. The propagation is represented by a translation matrix

$$M_t = \begin{bmatrix} 1 & d \\ 0 & 1 \end{bmatrix} \qquad (6.1)$$

The action of a lens of focal length, f, is represented by a refraction matrix

$$M_r = \begin{bmatrix} 1 & 0 \\ -\frac{1}{f} & 1 \end{bmatrix} \qquad (6.2)$$

For example, when a lens of a specific focal length must be used (see Section 6.6), or when the overall length of the Fourier optics system must be reduced [65], the Fourier lens must be replaced by a pair of lenses, f_1 and f_2, separated by a distance, d. In this case, the resultant propagation matrix is given by

$$\begin{bmatrix} A & B \\ C & D \end{bmatrix} = \begin{bmatrix} 1 & 0 \\ -\frac{1}{f_2} & 1 \end{bmatrix} \begin{bmatrix} 1 & d \\ 0 & 1 \end{bmatrix} \begin{bmatrix} 1 & 0 \\ -\frac{1}{f_1} & 1 \end{bmatrix} = \begin{bmatrix} 1 - \frac{d}{f_1} & d \\ -\frac{1}{f_2} + \frac{d}{f_1 f_2} - \frac{1}{f_1} & 1 - \frac{d}{f_2} \end{bmatrix}$$

The focal length of the lens combination is given by $-1/C$, which is

$$f = \frac{f_1 f_2}{f_1 + f_2 - d} \qquad (6.3)$$

This formula allows the construction of correlators employing two SLMs, where the focal length required does not correspond to that available from stock optics. These systems are discussed in Section 6.6.

When neither an achromat nor a combination of lenses suffice, the Fourier lens should be designed from scratch. There are a number of papers which indicate the approach to be adopted, some of which were mentioned in Section 1.3.1. The papers presented provide the experience of past workers who were concerned with the design of relatively long focal length Fourier lenses. The clearest exposition of the design problem is given in [22], where three-lens and four-lens designs are presented. These designs give a small amount of negative (barrel) distortion. This is not important in systems where the Fourier plane filtering is performed by a filter which has been fabricated using the same Fourier lens (Section 6.4). However, in systems where the filtering is performed by an SLM (Section 6.6), the spatial frequencies should be linearly spaced in the frequency plane. The spatial frequencies diffracted by either a transparency or an SLM when it is illuminated with coherent light are proportional to the sine of the diffracted angle. When there is no distortion in the lens, e.g. a camera lens, a collimated beam is focussed to a spot in the focal plane on the product of the focal length of the lens and the tangent of the incident angle of the beam. In order to space the spatial frequencies proportional to the sine of the diffracted angle, a small amount of barrel distortion must be present. Since a camera lens will image the diffracted spot at $f\tan\theta$ instead of $f\sin\theta$, the performance limit of the camera lens arises when the difference between $f\tan\theta$ and $f\sin\theta$ is equal to the radius of the diffraction limited spot [270]. If the Sony 4K SXRD is used in the aperture plane of the system and its aperture is filled, an elliptical diffracted spot results with minor and major axes equal to $\lambda f/34.8$ and $\lambda f/18.3$. In the case of a wavelength of 650 nm, the minor axis of the spot limits the diffracted angle to around 1.9 degrees before the distortion is significant. Since the maximum diffracted angle of this SLM at 650 nm is 2.19 degrees, the Fourier lens should include a small amount of barrel distortion in order to produce a linear spread of the higher spatial frequencies. This assumes that a diffraction limited spot can be achieved over the aperture of the diffracted field, which will be true for a well-corrected Fourier lens. A well-corrected Fourier lens contains up to 5 elements. This presents demanding assembly issues, which can be ameliorated by tolerancing the lens. Reducing the number of elements by, for example, adopting a converging beam arrangement (Figure 6.1b) can be sometimes beneficial [24, 125].

6.2.2 Diffraction integrals

In order to calculate the output from an optical system composed of several planes between which diffraction takes place, it is important to use a shorthand notation for the diffraction integrals. Vander Lugt realised the complexity of the mathematics required, and developed an appropriate shorthand notation [167]. The operator notation has been re-worked by several authors [37, 142, 187]. The formalism of [187] defines four operators. The one-dimensional correlates of these are multiplication by a quadratic phase exponential, $Q[c]$, and its inverse,

$Q[-c]$; the scaling by a constant, $V[b]$, and its inverse, $V[1/b]$; the Fourier transformation, $F[U(x)]$, and its inverse, $F^{-1}[\hat{U}(f_x)]$; and free-space propagation, $R[d]$, and its inverse, $R[-d]$ [90]. They are defined as follows:

$$Q[c](U(x)) = e^{\frac{ikcx^2}{2}}U(x); \tag{6.4}$$

$$V[b](U(x)) = \sqrt{|b|}U(bx); \tag{6.5}$$

$$F[U(x)] = \int_{-\infty}^{\infty} U(x)exp[-2\pi i f_x x]dx; \tag{6.6}$$

and

$$R[d][U(x)] = \frac{1}{\sqrt{i\lambda d}}\int_{-\infty}^{\infty} U(x)e^{\frac{ik}{2d}(x-x')^2}dx. \tag{6.7}$$

where the complex prefactor of the integral in Equation (6.6) is omitted, and the propagation in Equation (6.7) is between coordinates x and x' in planes separated by a horizontal distance of d.

The conventional arrangement for performing a Fourier transform is to place the input, $U(x,y)$, at the focal distance from a Fourier lens. This is called the front focal plane (ffp). When this input is illuminated with a collimated beam, which is a beam which forms a focus at infinity, then the Fourier transform, $\hat{U}(f_x)$, is formed at the focal distance behind the lens, which is the back focal plane (bfp). Two alternative arrangements are used when it is desired to shorten the track of the system (Figure 6.1a) and vary the scale of the Fourier transform (Figure 6.1b).

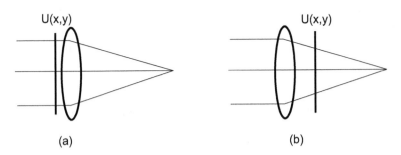

(a)   (b)

Figure 6.1 Two optical Fourier transform arrangements: (a) Input placed against the lens; (b) Converging beam

The application of the operator notation will be exemplified by using the Nazarathy and Shamir notation in the first case (Figure 6.1a) and the Vander Lugt notation in the second case (Figure 6.1b). The

Nazarathy and Shamir notation for the arrangement of Figure 6.1(a), gives

$$U(x') = R[f]Q[-\frac{1}{f}]U(x); \qquad (6.8)$$

where $U(x')$ is the output at the bfp of the lens, $R[f]$ is the free space propagation through a distance f, and $Q[-\frac{1}{f}]$ is the quadratic phase exponential of the lens. Therefore,

$$
\begin{aligned}
U(x') &= \frac{1}{\sqrt{i\lambda f}} \int_{-\infty}^{\infty} e^{-\frac{ikx^2}{2f}} U(x) e^{\frac{ik}{2f}(x-x')^2} dx \\
&= \frac{1}{\sqrt{i\lambda f}} \int_{-\infty}^{\infty} U(x) e^{\frac{ikx'^2}{2f}} e^{-\frac{ikx'x}{f}} dx \\
&= \frac{1}{\sqrt{i\lambda f}} e^{\frac{ikx'^2}{2f}} \int_{-\infty}^{\infty} U(x) e^{-\frac{ikx'x}{f}} dx
\end{aligned}
$$

Therefore, the Fourier transform of $U(x)$ appears in the bfp of the lens, multiplied by a phase factor which is quadratic in the x' coordinate. This phase factor is not important if the focal plane is the last plane in the optical system, e.g., where an image sensor is placed. However, if it is an intermediate plane, then this phase factor can influence the quality of the final stage of the system. When the input is placed in the ffp of the lens, then this phase factor disappears. The latter is the recommended arrangement for the first and intermediate stages of a Fourier optical system.

Vander Lugt based his operational notation on Fresnel functions which he termed ψ functions. Their use can be illustrated in the case of the converging beam arrangement (Figure 6.1b). The ψ function is defined as

$$\psi(x, y; 1/d) = e^{\frac{ik(x^2+y^2)}{2d}}. \qquad (6.9)$$

Therefore, a lens of focal length, f, is represented by

$$\psi(x, y; -1/f) = e^{-\frac{ik(x^2+y^2)}{2f}}. \qquad (6.10)$$

Two important properties of Fresnel functions can be readily verified:

$$\psi(x, y; 1/d_1)\psi(x, y; 1/d_2) = \psi(x, y; 1/d_1 + 1/d_2), \qquad (6.11)$$

and

$$\psi(x - x', y - y'; 1/d) = \psi(x, y; 1/d)\psi(x', y'; 1/d)e^{-\frac{ik(xx'+yy')}{d}}. \qquad (6.12)$$

The last property is important to describe a light beam propagated between two planes, (x, y) and (x', y'), separated by a distance d:

$$O(x', y') = \frac{c}{d} \iint_{-\infty}^{\infty} I(x, y)\psi(x - x', y - y'; 1/d)\mathrm{d}x\mathrm{d}y, \qquad (6.13)$$

where c is a constant of dimension L^{-1}, which will be replaced by unity from now on. This is a particular case of Equation 4.2 where the light is monochromatic and there are no aberrations.

The FT of a Fresnel function is another Fresnel function

$$g(x', y') = \iint_{-\infty}^{\infty} \psi(x, y; 1/d_1)e^{-\frac{ik(xx' + yy')}{d_2}} \mathrm{d}x\mathrm{d}y \qquad (6.14)$$

$$= bd_1\psi(x', y'; -d_1/d_2^2). \qquad (6.15)$$

where b is a complex constant of dimension L.

In reference to Figure 6.1(b), using (x'', y'') for the coordinates of the lens plane, and d_1, d_2 for the distances from the lens to the input plane and from the input plane to the output plane, respectively, then

$$U(x', y') = \frac{1}{d_1 d_2} \iint_{-\infty}^{\infty} \iint_{-\infty}^{\infty} \psi(x'', y''; -1/f)\psi(x'' - x, y'' - y; 1/d_1).$$
$$f(x, y)\psi(x - x', y - y'; 1/d_2)\mathrm{d}x''\mathrm{d}y''\mathrm{d}x\mathrm{d}y$$

$$U(x', y') = \frac{1}{d_1 d_2} \iint_{-\infty}^{\infty} \iint_{-\infty}^{\infty} \psi(x'', y''; -1/f)\psi(x'', y''; 1/d_1).$$
$$\psi(x, y; 1/d_1)e^{\frac{-ik(xx'' + yy'')}{d_1}} f(x, y)\psi(x, y; 1/d_2).$$
$$\psi(x', y'; 1/d_2)e^{\frac{-ik(xx' + yy')}{d_2}} \mathrm{d}x''\mathrm{d}y''\mathrm{d}x\mathrm{d}y$$

$$U(x', y') = \frac{1}{d_1 d_2}\psi(x', y'; 1/d_2) \iint_{-\infty}^{\infty} \iint_{-\infty}^{\infty} \psi(x'', y''; 1/d_1 - 1/f).$$
$$e^{\frac{-ik(xx'' + yy'')}{d_1}} \psi(x, y; 1/d_1 + 1/d_2)f(x, y)e^{\frac{-ik(xx' + yy')}{d_2}} \mathrm{d}x''\mathrm{d}y''\mathrm{d}x\mathrm{d}y$$

$$U(x', y') = \frac{1}{d_2}\psi(x', y'; 1/d_2) \iint_{-\infty}^{\infty} \psi(x, y; -f/d_1(f - d_1)).$$
$$\psi(x, y; 1/d_1 + 1/d_2)f(x, y)e^{\frac{-ik(xx' + yy')}{d_2}} \mathrm{d}x\mathrm{d}y$$

The Fresnel functions in the integral can be combined as follows

$$\psi(x, y; -f/d_1(f - d_1))\psi(x, y; 1/d_1 + 1/d_2)$$
$$= \psi(x, y; -f/d_1(f - d_1) + 1/d_1 + 1/d_2)$$
$$= \psi(x, y; -d_1(d_1 + d_2) + d_1 f)/d_1 d_2(f - d_1).$$

Since $f = d_1 + d_2$, the Fresnel function is equal to zero. Therefore,

$$U(x', y') = \frac{1}{d_2}\psi(x', y'; 1/d_2) \int\int_{-\infty}^{\infty} f(x, y)e^{\frac{-ik(xx'+yy')}{d_2}} \, \mathrm{d}x\mathrm{d}y \quad (6.16)$$

This is a scaled FT with a quadratic phase factor. The scale of the FT is reduced as d_2 is reduced.

6.3   HOLOGRAPHIC RECORDING MATERIALS

In a similar way to the development of SLMs, holographic recording materials (HRMs) have been developed for a different application area, in this case holographic memory. The resolution in each of the preceding HRMs is higher than the wavelength of visible light. This allows high density recording and a good range of angles for the recording beams. This makes the HRM ideal for high density, volume holographic storage. Although a large variety of storage media have been researched [242], it is silver halide emulsion, photopolymers, and photorefractive materials, which will be discussed here. The modern equivalent of the red sensitive silver halide emulsion that was used by Vander Lugt is the Slavich PFG-01 emulsion with a median grain size of 40 nm or the PFG03M with a median grain size of 10 nm. The resolution of such material far exceeds the requirements of off-axis holography. The layer thickness is around 7 microns, with a resolution of 3000 mm^{-1}, and sensitivity of 80 $\mu J/cm^2$. The maximum sensitivity of this emulsion is at 633 nm. The main disadvantage with an emulsion is that the plate must be removed from the system in order to develop the hologram. However, while it is removed from the system, the plate can be bleached in order to convert the amplitude hologram to a phase hologram [242]. This improves the diffraction efficiency (Chapter 3). In order to re-place the developed hologram in the system in a reproducible manner, a kinematically registered holder is employed [247]. Kinematic registration ensures the correct registration in the six degrees of freedom corresponding to the x-, y-, and z- axes, and rotations about the same axes.

The former use of silver halide emulsions has been replaced by the use of photopolymers, which have been found to be very promising because of their good resolution, high diffraction efficiency (DE), and real-time dry processing [27]. They were not mentioned 40 years ago, in the standard reference in this field [242]. However, the intervening period has seen an intensive research period in data storage, and the current growth in security applications has fuelled photopolymer development. The main components of the material are a polymerizable monomer, photoinitiator, and a sensitizer. Spatial variations in incident light intensity are recorded via irreversible changes in the refractive index. This produces a phase modulation, with its higher attendant diffraction efficiency. This change in refractive index is a result of photopolymerization reactions occurring in the bright regions of the incident interference pattern. Polymerization also results in development of monomer concentration gradient between the bright and dark zones, which promotes migration of monomer from dark to bright zones, thereby further enhancing the refractive index contrast. It is this refractive index modulation between the bright and dark regions that encodes the intensity modulation of the incident light. A variety of photopolymers have been reported, which are suitable for recording phase holograms. In particular, Bayfol®HX 102, has a 16.8 micron photopolymer layer with a dye to promote absorption at 633 nm, where the maximum index modulation is 0.03. Unfortunately, it cannot be used for wavelengths beyond 660 nm.

Photorefractive (PR) crystals have a longer history than photopolymers for holographic storage. The "optical damage", which is the basis of the photorefractive effect, was discovered in lithium niobate in 1968 [242]. During the past 50 years, the PR effect has been extensively studied for holographic storage and many other materials have been discovered, and polymers which exhibit this effect have been synthesized. The photorefractive effect is associated with crystals or polymers that exhibit a significant linear electrooptic effect. Upon irradiation, electrons are released from impurities by photoionization. When the irradiation is spatially inhomogeneous, the electrons migrate from the areas of high irradiance to those of low irradiance. This migration is a consequence of diffusion. The diffusion proceeds until the internal field generated by the migrated charge is of sufficient strength to generate an opposing electric field. It is this internal field which is responsible for the refractive index modulation, via the linear electrooptic effect. The diffusion length is commonly limited by trapping, so that an electric field is applied in order to increase the diffusion length to the fringe spacing of the pattern which is being recorded. The diffraction efficiency of a recorded grating increases in proportion to the square of the applied field. The provision of good quality crystals is reliant on experienced crystal growers, and generally the cost of these materials is relatively high. Therefore, recent research has seen the synthesis of photorefractive polymers as a replacement for the crystals. However, the dependence on relatively high electric fields remains an issue for system integrators.

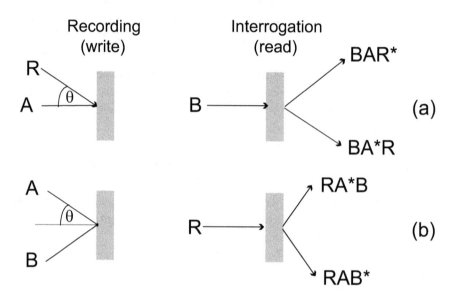

Figure 6.2 Recording geometries: (a) VLC; (b) JTC

In the correlators which will be discussed in Sections 6.4 and 6.5, there are two geometries for writing and reading the holograms. These are illustrated in Figure 6.2. When a reference beam (R) is incident at an angle θ to a signal beam (A), an off-axis hologram is formed (Figure 6.2a). The reference beam is represented by Re^{irx}, where R is the uniform amplitude of the beam and r is the phase ramp $-k\sin\theta$. The resulting interference pattern is the squared modulus of the total complex amplitude

$$|U(x,y)|^2 = |Re^{irx} + A(x,y)|^2$$
$$= R^2 + |A(x,y)|^2 + Re^{irx}A^*(x,y) + Re^{-irx}A(x,y)$$

The complex amplitude of the signal beam can be written as $A(x,y) =$

$|A(x,y)|exp[i\phi(x,y)]$, so that

$$|U(x,y)|^2 = R^2 + |A(x,y)|^2 + RA(x,y)|e^{i(rx-\phi(x,y))}$$
$$+R|A(x,y)|e^{-i(rx-\phi(x,y))} = R^2 + |A(x,y)|^2$$
$$+2R|A(x,y)|cos(rx - \phi(x,y))$$

The design of HRMs entails establishing operating parameters for which the recording is proportional to the light intensity. Under these conditions, the resulting interferogram is a sum of two zero-order terms and a cosinusoidal term. The dominant spatial frequency in the cosinusoidal term is $sin\theta/\lambda$. Therefore, when this interferogram is interrogated with a normally incident signal beam (B), the transmitted light is the product of B with the intensity of the interferogram

$$B(x,y)|U(x,y)|^2 = B(x,y)R^2 + B(x,y)|A(x,y)|^2+$$
$$B(x,y)A^*(x,y)Re^{irx} + B(x,y)A(x,y)Re^{-irx}$$

(6.17)

This consists of three beams, the first two terms form an on-axis beam, and the third and fourth terms give off-axis beams at angles θ and $-\theta$ to the normal to the HRM (Figure 6.2a).

When two signal beams (A) and (B) are incident at angles θ and $-\theta$ to the normal to the HRM (Figure 6.2b), the resulting interference pattern is

$$|U(x,y)|^2 = |A(x,y)e^{irx} + B(x,y)e^{-irx}|^2 = |A(x,y)|^2 + |B(x,y)|^2$$
$$+ A(x,y)B^*(x,y)e^{2irx} + A^*(x,y)B(x,y)e^{-2irx}$$

(6.18)

If the filter image beam is written as $A(x,y) = |A(x,y)|exp[i\phi(x,y)]$ and the input image beam is written as $B(x,y) = |B(x,y)|exp[i\psi(x,y)]$, then

$$|U(x,y)|^2 = |A(x,y)|^2 + |B(x,y)|^2$$
$$+ |A(x,y)||B(x,y)|e^{i(2rx+\phi(x,y)-\psi(x,y))}$$
$$+ |A(x,y)||B(x,y)|e^{-i(2rx+\phi(x,y)-\psi(x,y))}$$
$$= |A(x,y)|^2 + |B(x,y)|^2$$
$$+ 2|A(x,y)||B(x,y)|cos(2rx + \phi(x,y) - \psi(x,y))$$

The resulting interferogram is a sum of two zero-order terms and a cosinusoidal term. The dominant spatial frequency in the cosinusoidal term is $2sin\theta/\lambda$. When this interferogram is interrogated with a normally incident reference beam (R), the transmitted light is the product of R with the intensity of the interferogram

$$R|U(x,y)|^2 = R|A(x,y)|^2 + R|B(x,y)|^2 + RA(x,y)B^*(x,y)e^{2irx}$$
$$+ RA^*(x,y)B(x,y)e^{-2irx}$$

$$(6.19)$$

This consists of three beams, the first two terms form an on-axis beam, and the third and fourth terms give off-axis beams at angles 2θ and -2θ to the normal to the HRM (Figure 6.2b).

The development of HRMs has been similar to SLMs in a major respect, they have been developed for a different application area, in this case holographic memory. The resolution of each of the preceding HRMs is higher than the wavelength of visible light. This allows high density recording and a good range of angles for the recording beams. Moreover, an enhanced diffraction efficiency is achieved when the depth of the recording is greater than approximately $\lambda/sin^2\theta$. Under this condition, the grating is considered "thick" [242]. In addition to high diffraction efficiency, the angular sensitivity is high, so that small deviations of the angle of the illumination beam reduce the diffraction efficiency considerably. This makes the HRM suitable for high density, volume holographic recording, because a large number of holograms can be recorded with a very small crosstalk. However, when the same materials are used for Fourier plane filtering in an optical correlator, this angular sensitivity results in a decrease in intensity of the correlation beam as the object is displaced to different spatial positions in the input scene.

It was demonstrated in Section 1.3.3 that the correlation signal is displaced in proportion to the displacement of the input image. This mathematical feature of the correlation integral is known as shift invariance. The ideal Fourier plane filter would maintain the intensity of the correlation signal as the target moves across the input plane. This is a property of a space invariant interconnection, as discussed in Chapter 3. One of the factors which can limit the space invariance is the depth of the recording. When the depth of the recording is greater than approximately $\lambda/sin^2\theta$, the grating is considered "thick" [242]. In this case, the angular sensitivity is high, so that small deviations of the illumination beam reduce the diffraction efficiency considerably. Thick gratings produce a correlation signal intensity which is dependent on the object position [66]. The interconnection between the input plane and the correlation plane now becomes spatially variant in that the transfer function changes as the object is displaced across the input plane. Some workers, however, prefer to distinguish between this effect and spatial variance due to system level limitations [31].

6.4 THE VANDER LUGT CORRELATOR

6.4.1 Background

Gabor demonstrated an interference diagram between a point reference and a 1.4 mm diameter photograph in 1948, using a mercury-vapour lamp and pinhole to create the point reference [85]. The linewidth and spatial coherence of the source were adequate to allow a reconstruction of the original photograph from the photographic diagram. Moreover, it was clear that a 3D object could be recorded by this means. This was the beginning of the field of holography. The Willow Run Laboratories of the University of Michigan was one of many groups which started investigating this new field. When A. B. Vander Lugt joined in 1959, his colleagues were investigating how Fourier optical systems could be used to obtain maps from synthetic-aperture, side-looking radar systems. This work led to the invention of the off-axis image hologram [155]. The challenge for Vander Lugt was the fabrication of a complex-valued spatial filter for an optical correlator which is matched in both amplitude and phase to the spectrum of the input image. In order to achieve this task, he had to fabricate a holographic spatial filter by interfering the complex-valued spatial filter with a reference beam. After some initial abortive attempts using a high pressure mercury arc bulb with a bandpass filter, he was fortunate to be able to replace this source with one of the first helium-neon gas lasers delivered by Spectra-Physics. This laser was invented in 1960, and commercial fabrication began shortly afterwards. He was also fortunate to have available the new photographic film that was sensitive to the red light produced by these lasers. In early December 1962, he completed the development of the off-axis holographic spatial filter (HSF). The architecture based on the HSF came to be known as the Vander Lugt correlator (VLC) [264].

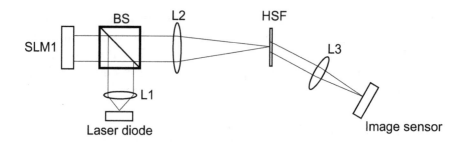

Figure 6.3   VLC employing an HSF. BS-Beamsplitter; HSF-Holographic spatial filter; L-Lens; SLM-Spatial light modulator

6.4.2   Original VLC

The original VLC was composed of a photographic transparency of the input image, a Fourier lens, a transparency of the holographic spatial filter, a second Fourier lens, and an output image sensor. It is sometimes called the 4f correlator because the spacing between each of these 5 components is f, the focal length of the lenses which are used. In Figure 6.3, the transparency has been replaced by an SLM input, SLM1, and the gas laser has been replaced by a diode laser and collimation lens. When SLM1 is illuminated by a collimated laser beam, the first lens forms a FT of the input image at the Fourier plane, where the HSF is located.

The HSF is the interference pattern between a collimated, off-axis reference beam and the complex conjugate of the FT of the template. The interference pattern, or interferogram, is prepared in a separate experiment (Figure 6.4). A major practical issue with this interferogram is the selection of the beam intensity ratio between the template FT and the reference beam. A FT has a high dynamic range, where the low spatial frequencies have a high amplitude and the higher spatial frequencies have a low amplitude. If the reference beam intensity is selected so that the beam intensity ratio is unity for the low spatial frequencies, then high contrast fringes will be formed at these frequencies. High contrast fringes of large magnitude are ideal for high diffraction efficiency when the HSF is used in the VLC. However, the filter will lack specificity, so that any object with a similar low-frequency spectrum will generate a correlation signal. When the intensity of the reference beam is selected so that the beam intensity ratio is unity for the high spatial frequencies, then high contrast fringes will be formed at these frequencies. Since the intensity at these frequencies is low, the fringes will have a low magnitude and a correspondingly low diffraction efficiency. However, the filter will have a high specificity because the high spatial frequencies are a signature of the particular object which is being sought. This is known as a weighted filter, where weighting is the process of emphasizing a particular range of spatial frequencies.

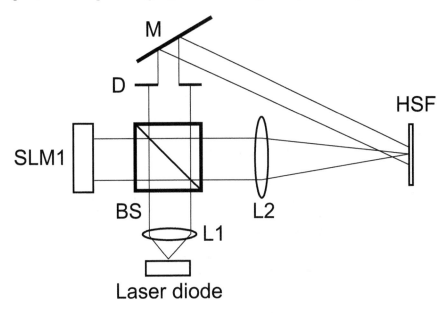

Figure 6.4 System for fabricating the HSF. BS-Beamsplitter; D-Diaphragm; HSF-Holographic spatial filter; L-Lens; M-Mirror; SLM-Spatial light modulator

When the FT of the input image is incident on this recording, the transmitted light is the product of the complex amplitude of the FT and the intensity of the interferogram (Equation 6.17). The transmitted light separates into three beams: a beam composed of the first two terms which remains on the optical axis; and two beams which separate as in Figure 6.2(a). The beam on the optical axis is known as the zero order intensity. The FT of the third term produces a correlation of the image and the template (Equation 1.14), and the FT of the fourth term produces their convolution (Equation 1.11). The phase ramps attached to these two terms give displacements of rf/k and $-rf/k$ in the output plane, respectively (Figure 6.2a). The displacements allow the measurement of either of these terms which are of small magnitude compared with the zero order. The zero order extends to approximately twice the width of the filter image, due to the second term in the equation. Therefore, the phase ramp should be designed to provide an adequate angular displacement of the three beams.

The layout of the VLC shown in Figure 6.4 corresponds to that with a reflective SLM, as used in [247]. The original VLC employed transmissive photographic transparencies [264]. This is called a matched filter (MF) and will be discussed along with other filters in the next chapter. The optical performance of the filter can be improved in two ways. The FT of an object transparency is generally of high dynamic range, and it is not straightforward to select the intensity of the reference beam. If the plane of the recording is displaced from the Fourier plane, then the dynamic range can be reduced [52].

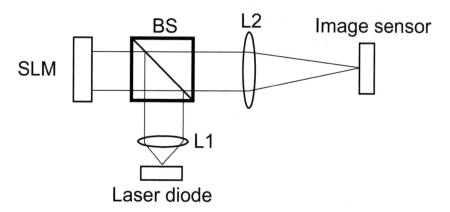

Figure 6.5 System for recording the Fourier transform of the pattern on the SLM on an image sensor. BS-Beamsplitter; L-Lens; SLM-Spatial light modulator

6.5   THE JOINT TRANSFORM CORRELATOR

The VLC consists of two separate planes containing an input image, on the one hand, and a filter, on the other hand. This is known as fre-

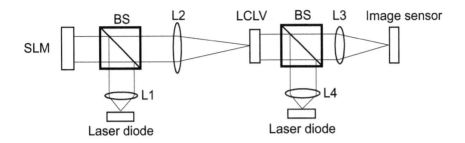

Figure 6.6   JTC recording the interference of the pattern on the SLM on an LCLV. BS-Beamsplitter; L-Lens; LCLV-Liquid crystal light valve; SLM-Spatial light modulator

quency domain filtering. The joint transform correlator (JTC) where the filter is Fourier transformed and placed adjacent to the input image, was devised as an alternative to the VLC [280]. This is known as spatial domain filtering. Both the input image and the spatial domain representation of the filter form a pattern which is Fourier transformed by a common Fourier lens and the interferogram of their FTs is recorded in the Fourier plane, on either an image sensor (Figure 6.5), an HRM, or a LCLV (Figure 6.6).

Suppose that the input image is a CxD matrix of values, with individual elements $I(c\Delta x, d\Delta y)$, and the filter is a ExG matrix of values, with individual elements $F(e\Delta x, g\Delta y)$. Their centres are separated by H pixels in the x direction, aligned in the y direction, and all pixels not occupied by the matrices are zero. Then the input pattern is

$$U(x, y, 0) = I(x - H\Delta x/2, y) + F(x + H\Delta x/2, y) \qquad (6.20)$$

The Fourier transform of a centred input image is (Equation (2.4))

$$\hat{I}(f_x, f_y) = \sum_{c=0}^{C-1} \sum_{d=0}^{D-1} I(c\Delta x, d\Delta y) exp\left[-2\pi i\left(f_x c\Delta x + f_y d\Delta y\right)\right] \qquad (6.21)$$

and the Fourier transform of a centred filter is

$$\hat{F}(f_x, f_y) = \sum_{e=0}^{E-1} \sum_{g=0}^{G-1} F(e\Delta x, g\Delta y) exp\left[-2\pi i\left(f_x e\Delta x + f_y g\Delta y\right)\right]$$

$$(6.22)$$

The displacements of the input image and filter from the optical axis produce linear phase ramps on the FTs of these distributions in the Fourier plane. This is a consequence of the Fourier shift theorem (Section 1.3.4), where a phase ramp of r in the input plane produced a displacement of $-rz_0/k$ in the FT. Here, a displacement of $-H\Delta x/2$ in the input plane produces a phase ramp of $H\Delta xk/2z_0$ in the FT. Similarly, a displacement of $H\Delta x/2$ produces the opposite phase ramp of $-H\Delta xk/2z_0$ in the FT. The intensity distribution in the plane of the image sensor or LCLV will be

$$|U(x',y',z_0)|^2 = \frac{1}{(\lambda z_0)^2}[|\hat{I}|^2 + |\hat{F}|^2 + \hat{I}\hat{F}^* exp(-2\pi i f_x H\Delta x)$$
$$+ \hat{I}^*\hat{F} exp(2\pi i f_x H\Delta x)] \quad (6.23)$$

This is called the Joint Transform Power Spectrum, or Joint Power Spectrum (JPS), and is of the same form as Equation (6.18). The interferogram recorded on the image sensor in Figure 6.5 can be transferred directly to the SLM of the same system as a complex amplitude distribution

$$U_2(x,y,0) = \frac{1}{(\lambda z_0)^2}[|\hat{I}|^2 + |\hat{F}|^2 + \hat{I}\hat{F}^* exp(-2\pi i f_x H\Delta x)$$
$$+ \hat{I}^*\hat{F} exp(2\pi i f_x H\Delta x)] \quad (6.24)$$

where $U_2(x,y,0)$ is the complex amplitude of the interferogram. When the SLM is interrogated with a normally incident reference beam (R), this gives four terms in a similar manner to Equation (6.19). The first two terms remain on the optical axis and are, respectively, the auto-correlation of the input image and the filter image. The significance of the phase ramps is that the FT separates the third and fourth terms from the first two terms. Both the third and fourth terms are correlations of the image and the filter function. The phase ramps attached to these two terms give displacements of $H\Delta x$ and $-H\Delta x$ in the output plane, respectively, because the focal length of lens L2 in Figure 6.5 is common to both the recording and interrogation systems. In the system of Figure 6.6, the focal length of L3 should be the same as L2 in order that the displacements on the image sensor correspond to the values given. The displacements allow the measurement of either of these terms which are of small magnitude compared with the zero order. In the JTC, the zero order extends to the sum of twice the widths of the input image and filter image. Since the input image is generally of larger SBWP than the filter image, the zero order in the JTC is twice the extent of the zero order in the VLC. This entails an adequate separation of the input and filter images in the input plane, which impacts the SBWP available for the input image.

The original advantage of the JTC was that it avoided the use of a recording medium to store the Fourier transform of the filter. This advantage has been eroded by the use of SLMs in the filter plane (Section 6.6). However, the capture of the JPS on an image sensor or LCLV allows Fourier plane processing which can improve the correlator performance. Binarisation of the JPS can be performed with a binary LCLV [120], and with an image sensor [122]. The nonlinear transfer characteristic of an LCLV can also be tailored to provide more information about the correlation spot [117].

6.6 HYBRID CORRELATORS

The first mention of a hybrid optical correlator was a name given to the JTC [47]. The original VLC stimulated the development of spatial light modulators (SLMs) which were used in place of transparencies in both the input and filter planes [207]. Where the introduction of an SLM in the input plane allowed digital pre-processing of the input image and also the introduction of images from a frame store, the system was described as hybrid [60]. Similarly, the application of digital algorithms to calculate the filter function when an SLM replaced the HSF in the filter plane of the VLC was described as a hybrid implementation [276]. An example of a two SLM VLC is given in Figure 6.7.

The light from a laser diode is collimated by a lens arrangement, L1, and is normally incident on SLM1 using a beamsplitter (BS). The light reflected from SLM1 and transmitted by the BS is incident on a Fourier lens, L2, which produces a FT of the input image on SLM2. When SLM2 displays the filter function, the reflected beam is modulated by the product of the FT of the image and the filter function. The portion of this beam reflected by a second BS is Fourier transformed by lens L3 and the correlation between the image and the FT of the filter function is recorded on the image sensor. A hybrid correlator employing two LCOS devices with a processing power of 0.6 Teraflops (10^{12} floating point operations per second) has been developed by Innovative Signal Analysis, Inc.

In order to fully utilise the SBWP of SLM2, the FT of the input image has to be scaled appropriately. This scaling is performed by

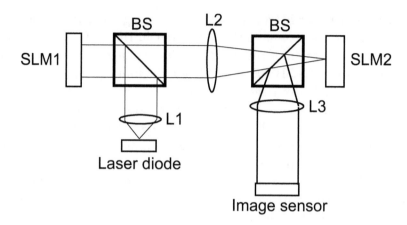

Figure 6.7 Hybrid correlator employing two EASLMs. BS-Beamsplitter; L-Lens; SLM-Spatial light modulator

adjusting the focal length of lens, L2. In order to calculate the focal length of L2, the size of the FT of the image is calculated and equated to the size of SLM2. For square SLMs, the highest spatial frequency generated by SLM1 should be located at the corner pixel of SLM2. When this condition holds, and SLM1 has the same resolution as SLM2, then every element of the FT in Equation (2.3) coincides with a pixel of SLM2. For SLMs of rectangular aspect ratio, the highest spatial frequency should match the minor axis of SLM2, and the pixels along this axis will be in coincidence with the elements of the digital FT along this dimension. For an SLM with a pixel repeat of Δx_1, the largest spatial frequency generated is $\eta_{max} = 1/2\Delta x_1$. In the focal plane of the Fourier lens, this is located at $x'_{max} = f\lambda/2\Delta x_1$. Therefore, if the number of pixels and the pixel repeat of SLM2 are L and Δx_2 respectively, then

$$\frac{f\lambda}{2\Delta x_1} = \frac{L\Delta x_2}{2} \tag{6.25}$$

This equation can be used to calculate the required focal length of the Fourier lens, according to

$$f = \frac{L\Delta x_2 \Delta x_1}{\lambda} \tag{6.26}$$

Two lenses, f_1 and f_2, separated by a distance, d, are used to synthesize

f, according to Equation (6.3)

$$f = \frac{f_1 f_2}{f_1 + f_2 - d} \tag{6.27}$$

The combination of two positive lenses were used for this purpose in the correlator constructed by the Naval Undersea Warfare Center, where two Forth DD SXGA-R2D 1280 x 1024 FLCOS screens were used for the input and template SLMs [173]. The combination of a positive lens and a negative lens is used in [231] and can be used to reduce the overall length of the correlator [65].

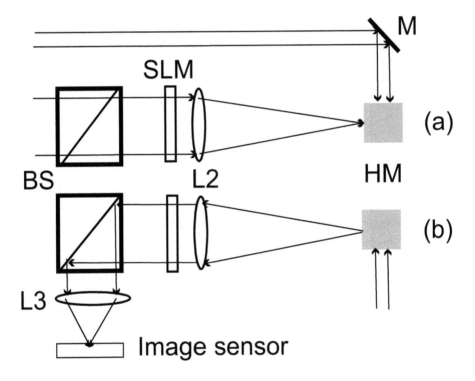

Figure 6.8 Inverted VLC: (a) storing the templates in an HM; (b) replay of the templates from the HM. BS-Beamsplitter; HM-Holographic memory; L-Lens; SLM-Spatial light modulator

A novel type of hybrid correlator employed electronic processing to form the 2D digital FT of the scene on a transmissive EASLM, which was optically addressed by THE FTs of the template originating in a holographic memory (HM) (Figure 6.8) [286]. The rationale for this approach was that, when a hybrid correlator is required to search through a number of filters for each input scene, the number of correlations per second (cps), is limited by the speed of the template SLM. The template SLM is replaced by an angle multiplexed HM, the access speed of which is much faster. The different templates in a filter bank can be accessed at a template rate of kHz when the addressing beam is modulated by an acoustooptic deflector. In the system illustrated in Figure 6.8, the templates are stored by interfering the template beam with a plane reference wave (Figure 6.8a). The SLM displays the phase-only FT of the template so that the template is recorded in an edge-enhanced format in the HM. The HM used was a PR crystal called Lithium Niobate. The phase-only FT employs the phase modulation discussed in Section 2.3.1. It provides a high efficiency light throughput, which will be quantified in the Chapter 7. In order to perform the correlation of multiple templates with a given scene image, the templates are accessed by angle-multiplexing a plane reference wave which is counterpropagating to the reference wave which recorded the edge-enhanced template (Figure 6.8b) [68, 69]. The FT of the template beam counterpropagates to the SLM which now displays an FT of the scene image. The product of the conjugate template and the scene image is Fourier transformed by lens L3 to give a correlation signal on the image sensor. This system was constructed under European funding (Brite) and the performance details are provided in Table 6.1 [235].

The inverted (or transposed) VLC correlator is capable of searching through a significant number of templates for every input image. Therefore, it solves the problem of orientation and scale inflexibility in the standard correlator by allowing the storage of all the necessary orientational and scale changes in the template. A transposed correlator was the basis of a Small Business Innovation Research (SBIR) grant in the mid-1990s to Advanced Optical Systems, Inc. to develop a missile guidance system [35]. If, in addition, the input scene is of low SBWP, then the electronic processing can be significantly extended. The FFT of the input scene and the multiplication with the FT of the template can be performed electronically. This is most efficiently performed with binary versions of the two FTs. In which case the multiplication is an exclusive-OR (XOR) logic operation. The final FT is performed at high speed using a binary SLM [20, 178].

The systems shown in Figures 6.7 and 6.8 employ on-axis filtering, as opposed to the off-axis filtering used in the original VLC. The use of the off-axis reference beam provided the advantage of separating the three beams in the correlation output Equation (6.17). However, it required a high-resolution recording medium to encode the interference pattern. The SLMs used in these HCs do not have a large SBWP in comparison with the HRM. Therefore, they encode on-axis filters and the three beams are superposed in the output plane. Where it is important to separate the correlation output from the other outputs, the filter is displaced to the corner of the SLM, say in the second quadrant. Then the three beams will appear in the output plane, one on-axis, one in the second quadrant, and one in the fourth quadrant.

6.7   CORRELATOR SYSTEMS

During the 1990s many establishments constructed optical correlators. A list of correlator systems together with their principal characteristics is given in Table 6.1, where the hybrid correlators are indicated by HC. There were many proof-of-principle systems built on optical tables, but the ones listed in Table 6.1 were attempts to make a self-contained system which could be moved around and, in one case, taken on a car journey [96]. For example, the miniaturized optical correlator (MOC) was the result of a program established by the U.S. Army Missile Command to design field operable hardware. The correlator had to be miniaturized to fit into the seeker head of a 150 mm diameter test missile. This missile would free-fall from a height of 1525 m and the correlator had to provide guidance down to an altitude of 153 m [241]. A compact system based on earlier work [70] was delivered to the Jet Propulsion Laboratory (JPL). The input SLM was the Hughes LCLV (Section 2.6) with an associated telescope which imaged distant objects on the photosensor layer [247]. (The Hughes LCLV was commonly sold with a fibre optic faceplate which relayed the image from the front of the faceplate to the back surface, so that the telescope strictly imaged onto the front face of the faceplate.) Moreover, the optics required for collimating the read light on the LCLV was a folded optics assembly (Section 8.1.2). In order to improve the number of correlations per second, a ferroelectric liquid crystal SLM was used in the filter plane in the third generation miniature ruggedized optical correlator (MROC III). For faster speeds, the Brite correlator employed page-oriented holographic memory for the template storage. A high-speed acoustooptic deflector provides a random access time of 15 μs to this memory. The addressed template is Fourier transformed by lens L2 and incident on a TFT SLM, which holds the Fourier transform of the input scene (Figure 6.8). The light intensity diffracted from the holographic memory is low. Therefore, the filter SLM contains the template in phase modulated form, in order to promote light efficiency (Section 7.6.1). The JPL greyscale optical correlator (GOC) employed a fast ferroelectric greyscale SLM for the template images, which allowed a fast speed when searching through a large database of templates for each input image. Martin Marietta funded the construction of a fast correlator at Boulder Nonlinear Systems (BNS) [233]. Ferroelectric LCOS was used for both the input and template SLMs and this design was patented [239]. The work was pursued at BNS under funding from the U.S. Army Missile Command and resulted in an optical correlator equipped with 256 x 256 FLCOS operating at 4 kHz [74]. The overall speed of the system was limited by the camera head to 979 cps. The FLCOS were tested in analogue modulation mode, probably using ancillary retardation plates in front of the device [16]. The analogue mode results from intermediate orientations of the liquid crystal director between the two stable in-plane orientations discussed in Section 2.3.1. The retardation plates convert the partial rotation of the director into analogue phase values. Finally, the National Optics Institute of Canada (INO) produced an optical processor based on TFT LCDs which could be either configured as an HC with a phase modulated filter, or configured for spatial filtering (Section 8.5).

In Europe there was a complementary effort to build correlator systems funded mainly by the European Commission. The lead contractor in a European consortium, Thomson-CSF, constructed and successfully tested a compact correlator, which employed a PR crystal, bismuth silicon oxide (BSO), to record the JPS [210]. The PR crystal was complemented by a mini-YAG (yttrium aluminium garnet) laser and a TFT LCD input device. In the first version of this correlator, the speed was limited by the input device. In a later version, this was replaced by a 256x256 pixel FLC SLM, and gave in excess of 1000 cps [59]. The performance of this correlator for fingerprint recognition was documented in [218]. The European Space Agency funded TU Dresden to construct a number of correlators. The entry in Table 6.1 provides one example of these, based on a JTC with image sensor output which can be accommodated in a linear system of 22 cm length. The SBWP recorded is that of the input plane SLM. Where a JTC is implemented, the SBWP of the input image will be less than one half of the SBWP in the table, due to the need to accommodate both the input and template images, with appropriate spacing between them, on the same input device. The Institute of Optical Research (IOF) in Stockholm constructed a hybrid correlator which could be programmed remotely over an Internet connection [102]. The TFT devices used in the Brite, INO, and Thomson systems were liquid crystal television screens originally designed for projection systems, and adapted for use in correlator systems by adjusting the drive voltages.

In the Far East, the Japan Women's University has developed a number of correlators of which one, FARCO, is reported in the table.

6.8   3D CORRELATORS

Spatial light modulators are limited with regard to real world scenes. It was shown in Chapter 1 that the digital picture function is an approximation to the picture function of an object. However, when the object is embedded within a 3D scene, a number of complications arise. Scale, aspect, illumination, and occlusion of the object all have to be accommodated. In order to capture the 3D scene electronically for object recognition, four approaches will be mentioned: digital holography [121]; capturing a sequence of 2-D projections [219]; integral imaging [83]; and the construction of sub-images [197]. Digital holography is a technique for electronic processing of holograms in digital form. Typically, the hologram of a 3D scene is captured on an image sensor. This hologram is an interference pattern between the scene and a reference wave. Numerical reconstruction and cross-correlation with the template is performed electronically. Optical processing is beneficial when the reference template can be placed in the input scene [219]. It was demonstrated, by simulation of a JTC, that the 3D position of the object can be recovered. This involves the capture of 2D projections of the 3D scene by translating the image sensor in a plane orthogonal to the line connecting the viewer to the scene. In general, the imaging of 3D scenes is preceded by a camera calibration in order to calculate the elements of the homography matrix, which relates the coordinates of the image sensor to those of the scene, but, with suitable approximations in this case, the location of the template within the scene can be ascertained.

Integral imaging is more than 100 years old. It was invented in the form of integral photography by Gabriel Lippmann in 1908. Professor Lippmann was a prolific inventor who won the Nobel Prize in the same year for his invention of a method for colour photography based on recording interference fringes in a high-resolution silver halide emulsion. Integral photography involved the recording of a scene using an

TABLE 6.1 Compact correlator systems

Developer	System	Reference	Type	Input Plane	SBWP	Filter Plane	Speed (cps)
Brite	HDOC	[235]	HC	TFT	512x512	HM	3000
INO	OC-VGA3000	[18]	HC	TFT	512x480	TFT	30
IOF	IOF	[240]	HC	FLCOS	256x256	FLCOS	220
Japan Women's Uni.	FARCO	[277]	HC	FLCOS	1280x768	LCOS	1000
JPL	HDOCC	[231]	VLC	LCLV		LCLV	
JPL	MOC	[247]	VLC	LCLV	500x500	HSF	30
JPL	GOC	[55]	HC	TFT	640x480	FLCOS	10000
Litton Systems	MROC III	[41]	VLC	FLCOS	256x256	FLCOS	1920
Martin Marietta	SPOTR	[159, 233]	HC	FLCOS	128x128	FLCOS	500
Peugeot Citroen	HIPOCOS2	[96]	JTC	FLCD	128x128	BOASLM	400
Thomson	NAOPIA	[210]	JTC	TFT	320x264	BSO	60
TU Dresden	SMARTSCAN	[116]	JTC	TLCOS	320x240	CCD	77

array of small lenses in place of the single lens in a camera. When the photographic plate was developed and viewed with a similar array of lenses, the original scene exhibited parallax when viewed from different positions within the lens array aperture. These cues give a realistic rendition of the original 3D scene. Integral imaging is the name given to the first aspect of integral photography. It is the basis of light field cameras such as those marketed by Raytrix and Lytro. It is also known as plenoptic imaging. 3D object recognition based on integral imaging involves digital computation and is known as computational integral imaging [139]. The advantage of integral imaging compared with digital holography is that incoherent light can be used. An interesting use of lenslet arrays was the creation of sub-images from the elemental images formed by lenslet arrays [197]. The advantage of creating these sub-images is that they can be used for the measurement of out-of-plane rotations and shifts using sequences of 2D correlations. The disadvantage is that this involves the complexity of electronic wiring.

GLOSSARY

Holographic recording material (HRM): A medium which records high spatial frequencies by modification of its optical properties, especially the refractive index.

Hybrid correlator: An optical correlator which benefits from electronic processing to improve system performance.

Joint transform correlator (JTC): An optical correlator where the template is stored in electronic form and introduced alongside the input scene.

Joint power spectrum (JPS): The intensity profile of the Fourier transform of the input scene and template in a JTC.

Operator notation: A shorthand representation of common integrals appearing in diffraction calculations by operators.

Ray matrix: A concise representation of the effect of propagation and refraction on the ray height and angle in the paraxial (or Gaussian) approximation.

Transposed correlator: An optical correlator where an FT of the template illuminates the FT of the input scene.

Vander Lugt correlator (VLC): An optical correlator where the template is stored in permanent form on a recording medium.

Filtering

CONTENTS

7.1 SYNOPSIS

A fundamental task in image processing is that of detecting a relevant object in a scene, such as face recognition. If the position of the recognised face is required, then the task is called localisation. If a particular face is sought, then the task is called identification. In Chapter 1, the capability of digital cameras to recognise and localise faces was mentioned. Identification is a more complex task. This chapter is concerned with the engineering of filters in order to improve the performance of the correlator system in particular directions, such as: discrimination, in order to facilitate identification; throughput, in order to facilitate the detection task; and correlation peak sharpness in order to facilitate localisation. In the design of filters, the template of the object is described in vector format. Therefore, Section 7.2 will recap the relevant vector and matrix theory. The detection may be hampered by either background noise or variability of the template. Both noise and variable template are treated within the same mathematical framework which is the theory of random variables and functions. This framework will be addressed in Section 7.3. If the filter is designed in anticipation of these variabilities, then the object detection can be made more robust. The final decision on whether the object has been detected is discussed in the context of binary hypothesis testing in Section 7.4. For most of the correlator systems which were presented in Chapter 6, the template image was either the image which was sought, or the complex conjugate of its Fourier transform. In the latter case, the correlation operation is robust to added noise, and the signal to noise ratio (SNR) is optimal. The SNR of a filter is just one of a number of relevant figures of merit (FOMs) which are used for assessing the performance of the filter. These are discussed in Section 7.5. One of the important FOMs relates to the throughput efficiency of the filter. A significant development in optical correlators was the realisation that a phase-only version of the template was effective and had a high throughput efficiency. This is one of the filter coding techniques which are presented in Section 7.6. The phase-only filter is a good example of how one FOM, the filter efficiency, can be improved to the deficit of other FOMs, such as the discrimination capability (DC). A moderate degree of DC is important in order to distinguish the object from the non-object. However, small variations in the presentation of the object should be tolerated. Unfortunately, when a filter of high DC is employed, the correlation operation is no longer robust when the object in the scene differs slightly from the template image. The correlation output can vary significantly even for the same object when it is slightly altered, for example, by orientation angle. For example, when the input scene contains a rotated or scaled object, the correlation output can fall by 50% when the template is rotated by 0.2 degrees or scaled by 0.5% [46]. The elementary solution is to present a set of references which cover the variations in the best possible way. These can be presented either sequentially to the system (temporal multiplexing), or in parallel in a multichannel filter (spatial multiplexing), or combined into a composite filter. In order to design composite filters, a relevant set of reference images is collected. This is called the training set, and it covers the expected variations of the template. In addition to improving the robustness of the correlation for variations of the template, the filter design can also include an element which discriminates against non-target objects. Section 7.7 describes the design of these filters. The filters can be composed in either the space or frequency domain, and there can be advantages to one or the other approach. These filters are synthesized in vector form and then encoded onto a computer-generated hologram or a spatial light modulator. Finally, Section 7.8 discusses phase-only correlation (POC).

7.2   VECTORS AND MATRICES

A vector is a list of numbers and is written as a row list for the row vector or a column list for the column vector. The transpose operation converts the row vector to the column vector of the same numbers, and vice versa. A vector is denoted by a bold lowercase letter, so that, if $\mathbf{a}$ is a row vector, then $\mathbf{a}^T$ is the corresponding column vector. A matrix is an array of numbers which is denoted by a bold uppercase letter, so that, if $\mathbf{A}$ is a matrix composed of 3 rows and 2 columns (3 x 2 matrix), then $\mathbf{A}^T$ is a 2 x 3 matrix. Scene and template images in the spatial and frequency domains are stored digitally as matrices. They can also be represented as vectors for certain calculations, by raster scanning the matrix into a 1D column vector. For example,

$$\mathbf{A} = \begin{bmatrix} j & k \\ l & m \\ n & o \end{bmatrix} \tag{7.1}$$

can be represented by a column vector, $\mathbf{a}$, where

$$\mathbf{a} = \begin{bmatrix} j \\ k \\ l \\ m \\ n \\ o \end{bmatrix} \tag{7.2}$$

and the transpose of $\mathbf{a}$ is $\mathbf{a}^T$=[j k l m n o]. Similarly, the transpose of matrix $\mathbf{A}$ is the 2 x 3 matrix,

$$\mathbf{A}^T = \begin{bmatrix} j & l & n \\ k & m & o \end{bmatrix} \tag{7.3}$$

The product of a real row vector multiplied by a real column vector of the same number of components, $\mathbf{a}^T \mathbf{b}$, is a real number, which is known as the dot product, $\mathbf{a}.\mathbf{b}$. If the dot product is zero, then the vectors $\mathbf{a}$ and $\mathbf{b}$ are orthogonal. If, in addition, $\mathbf{a}.\mathbf{a}$ and $\mathbf{b}.\mathbf{b}$ are unity, then vectors $\mathbf{a}$ and $\mathbf{b}$ are orthonormal. The product of a column vector multiplied by a row vector, $\mathbf{a}\mathbf{b}^T$, is a matrix.

Since the FT of an image matrix is a complex matrix, the operations on complex matrices have to be considered. The conjugate transpose, which is the complex conjugate of the transposed vector, is denoted $\mathbf{a}^+$. The complex inner product of two vectors $\mathbf{a}$ and $\mathbf{b}$ is the product of every component of $\mathbf{a}^+$ multiplied by the corresponding component of $\mathbf{b}$, followed by summing the component products. The complex inner product is written as $\mathbf{a}^+ \mathbf{b}$.

The calculation of filters involves the product of rectangular matrices. The product of an N x M matrix multiplied by an M x N matrix is an N x N matrix. For example,

$$A^T A = \begin{bmatrix} j & l & n \\ k & m & o \end{bmatrix} \begin{bmatrix} j & k \\ l & m \\ n & o \end{bmatrix} = \begin{bmatrix} j^2 + l^2 + m^2 & jk + lm + no \\ kj + ml + on & k^2 + m^2 + o^2 \end{bmatrix} \quad (7.4)$$

and

$$AA^T = \begin{bmatrix} j & k \\ l & m \\ n & o \end{bmatrix} \begin{bmatrix} j & l & m \\ k & n & o \end{bmatrix} = \begin{bmatrix} j^2 + k^2 & jl + mk & jn + ok \\ lj + mk & l^2 + m^2 & ln + om \\ nj + ok & nl + om & n^2 + o^2 \end{bmatrix} \quad (7.5)$$

$A^T A$ is a 2 x 2 symmetric matrix with two components, $j^2 + l^2 + m^2$ and $k^2 + m^2 + o^2$ on the main diagonal. AA^T is a 3 x 3 symmetric matrix with three components, $j^2 + k^2$, $l^2 + m^2$, and $n^2 + o^2$, on the main diagonal. A symmetric matrix is one where the matrix transpose is equal to the matrix. Therefore, the transpose of a product of two matrices reverses the order of the matrices, $(AB)^T = B^T A^T$. The determinant of a square matrix is a sum of sub-products of the elements of the matrix. For example, $det(A^T A)$ or $|A^T A|$, is

$$\begin{vmatrix} j^2 + l^2 + m^2 & jk + lm + no \\ kj + ml + on & k^2 + m^2 + o^2 \end{vmatrix} = (j^2 + l^2 + m^2)(k^2 + m^2 + o^2)$$

$$-(jk + lm + no)^2$$

and det($\boldsymbol{AA}^T$) is

$$\begin{vmatrix} j^2 + k^2 & jl + mk & jn + ok \\ lj + mk & l^2 + m^2 & ln + om \\ nj + ok & nl + om & n^2 + o^2 \end{vmatrix} = (j^2 + k^2) \begin{vmatrix} l^2 + m^2 & ln + mo \\ nl + om & n^2 + o^2 \end{vmatrix}$$

$$- (jl + mk) \begin{vmatrix} lj + mk & ln + mo \\ nj + ok & n^2 + o^2 \end{vmatrix} + (jn + ok) \begin{vmatrix} lj + mk & l^2 + m^2 \\ nj + ok & nl + om \end{vmatrix}$$

Larger matrices follow a similar pattern, with a negative sign on the odd column sub-products.

Commonly, the optimum filter is an eigenvector of a matrix. The product of an N x N matrix, **B**, and an N-component vector, **b**, is an N-component vector, **c**. If $\mathbf{c} = m\mathbf{b}$, where m is a number, then **b** is called an eigenvector of the matrix and m is the associated eigenvalue. The eigenvectors and associated eigenvalues of a symmetric matrix can be found by matrix algebra. In the first place, the eigenvalues are found by solving the N simultaneous equations, $det(\boldsymbol{B} - m\boldsymbol{I}) = 0$, where **I** is the N-component identity matrix which has unity values for the main diagonal elements and zero elsewhere. These equations will have n solutions for m, $m_1, m_2, ... m_n, .. m_N$, when the eigenvectors are linearly independent. These are the N eigenvalues. Once the eigenvalues have been found, the eigenvector associated with each eigenvalue can be computed by solving the simultaneous equations $\boldsymbol{Bb} = m_n\boldsymbol{b}$. If the eigenvectors are ordered as the columns of a matrix **E** and the eigenvalues are correspondingly ordered in a diagonal matrix $\boldsymbol{\Lambda}$, then **B** can be expressed as the product of three matrices, $\boldsymbol{B} = \boldsymbol{E}\boldsymbol{\Lambda}\boldsymbol{E}^T$. This is known as the spectral decomposition of matrix **B**. When the matrix is rectangular, such as **A**, a similar factorization can be performed, $\boldsymbol{A} = \boldsymbol{Q}_1\boldsymbol{\Sigma}\boldsymbol{Q}_2^T$, where $\boldsymbol{Q}_1$ is a 3 x 3 matrix composed of columns which are eigenvectors of $\boldsymbol{AA}^T$, and $\boldsymbol{Q}_2$ is a 2 x 2 matrix composed of columns which are eigenvectors of $\boldsymbol{A}^T\boldsymbol{A}$. $\boldsymbol{\Sigma}$ is a 3 x 2 diagonal matrix with two non-zero components, which are the square roots of the two common eigenvalues of $\boldsymbol{A}^T\boldsymbol{A}$ and $\boldsymbol{AA}^T$. This is known as singular value decomposition (SVD) and software packages such as MATLAB can be used to find the factors. A symmetric matrix, where the matrix is equal to its transpose, has real eigenvalues and orthogonal eigenvectors. Orthogonality means that, for any vector, **b**, $\boldsymbol{b}^T\boldsymbol{b} = 0$. The inverse of a square matrix is a matrix, $\boldsymbol{D}^{-1}$, for which $\boldsymbol{DD}^{-1} = \boldsymbol{I}$.

In Section 1.4, the discrete Fourier transform (DFT) of a four component 1D vector was considered. The transform is a four component vector which is ordered from low to high spatial frequencies. The zero order component is the first component of the transform. The 2D DFT is ordered similarly, with the low spatial frequencies at the beginning of the rows and columns. In order to design filters for optical systems, where the zero order is at the centre of the matrix, a shift operation must be performed on the results calculated digitally. The zero order must be shifted from the top corner of the 2nd quadrant to the top corner of the 4th quadrant. In order to maintain the 1D transform nature of the rows and columns, the 1st and 3rd quadrants must be interchanged. This shift operation is shown for the 2D transform in Figure 7.1.

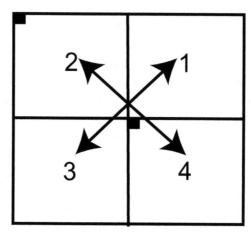

Figure 7.1   Shifting quadrants to convert a DFT matrix to a form suitable for an optical FT system

7.3   RANDOM VARIABLES AND FUNCTIONS

A random variable is a quantity which takes various values depending on the result of an observation, such as the number on the throw of a die. In this section, the statistical functions describing a single random variable, such as the noise superposed on a 2D scene, are presented initially. The mathematics developed can be extended to derive the random functions describing the variability of a template image, when a number of different manifestations of the template are available. These should encompass all variations of the object which are of interest. For example, when the object is someone's face, the face under varying illumination, pose, or expression should be considered.

When an input image is corrupted by noise, which is a random function of time, then the family of random variables which represents the noise is infinite, and the random function is called a random process. Most commonly, the random process is described by a cumulative distribution function (cdf), $F_t(T)$, which is the probability that the value of $n(t)$ is less than T. A multidimensional distribution function can also be defined for the multidimensional random variable, $n(t_1)$, $n(t_2)$,.....$n(t_m)$. If the points t_1, t_2........t_m represent different instances of time, and the points are shifted along the time axis, then a second random process is generated. If this second random process has an identical multidimensional distribution function to the first, then the random process is described as stationary. All noise processes considered here will be stationary. An example is additive white Gaussian noise. The probability density function (pdf) of this random variable, $p(x)$, is a Gaussian function,

$$p(x) = \frac{1}{\sigma\sqrt{2\pi}}e^{-\frac{(x-\mu)^2}{2\sigma^2}} \tag{7.6}$$

where μ is the mean and σ^2 is the variance. The mean value, μ, or expectation of $n(x)$ is defined by multiplying each x by the probability of its occurence and integrating

$$\mu = \int_{-\infty}^{\infty} xp(x)dx \tag{7.7}$$

The variance of $n(x)$ is the expectation of $n(x)^2$ centred by the mean value,

$$\sigma^2 = \int_{-\infty}^{\infty} (x-\mu)^2 p(x)dx \tag{7.8}$$

The pdf of a standard normal distribution, which is a Gaussian function with zero mean and unity variance, is plotted in Figure 7.2, together with its cdf. The cdf is the integral of this function between $-\infty$ and x. The variance is a measure of the spread of the distribution.

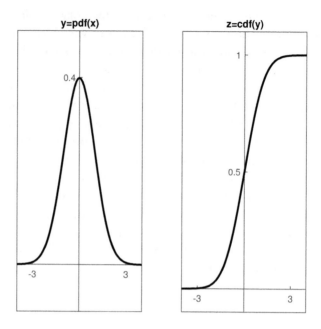

Figure 7.2 pdf of a standard normal distribution and its cdf

When Gaussian noise of zero mean and unity variance is superposed on the image, the numerical values at the pixel are selected at random from the pdf plotted in Figure 7.2. The 'x' values give the number which is added to the value of the image pixel and the corresponding 'y' value gives the relative frequency with which this 'x' value should be selected. The addition of the noise illustrated in Figure 7.2 to an image is shown in Figure 7.3. The noise has been multiplied (i.e., broadened) to different degrees by multiplying by a constant, which multiplies the variance by the same constant. Since the standard normal distribution is centred around zero, the additive noise can have both positive and negative values. Therefore, the resulting images have been normalised to values between 0 and 1.

Selecting two of the multidimensional random variables, $n(t_1)$, $n(t_2)$, then, in addition to the means, μ_1, μ_2, and variances, σ_1^2 and σ_2^2, of each random variable, the covariance of $n(t_1)$ and $n(t_2)$ expresses

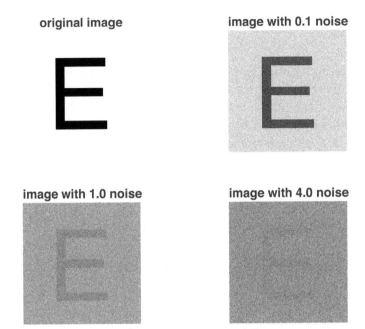

original image

image with 0.1 noise

image with 1.0 noise

image with 4.0 noise

Figure 7.3 Image with no noise and with added noise of variance 0.1, 1.0 and 4.0

the correlation of the variations of $n(t_1)$ and $n(t_2)$

$$Cov(n(t_1), n(t_2)) = \iint_{-\infty}^{\infty} (x - \mu_1)(y - \mu_2)p(x, y)dxdy \qquad (7.9)$$

where $p(x, y)$ is the joint probability of variable, $n(t_1)$, taking the value x, and variable, $n(t_2)$, taking the value y. For a stationary process, the covariance is a function of $\tau = t_1 - t_2$. Additive white Gaussian noise is a random signal with zero mean, a variance of σ^2, and a covariance of σ^2 if $\tau = 0$, and zero if $\tau \neq 0$.

The same statistical functions can be computed for random variables which are not infinite, using summations in place of integrations. If X and Y are discrete distributions of M components, they are represented as a column vector and

$$Cov(X, Y) = \sum_{l=1}^{M} \sum_{k=1}^{M} (x_l - m_X)(y_k - m_Y)^T \qquad (7.10)$$

where

$$m_X = \frac{1}{M} \sum_{l=1}^{M} x_l, \text{ and } m_Y = \frac{1}{M} \sum_{k=1}^{M} y_k. \qquad (7.11)$$

The same functions can be computed when there are more than two discrete distributions; for example, when a number of different manifestations of the template image are available. The N template images are expressed in raster scanned vector format, s_1, s_2,..s_n,...s_N, where each s_n is a column vector of M components, M being the number of pixels in the image. Then the mean, $\bar{s}$, of a set of N samples is the average over all samples

$$\bar{s} = \frac{1}{N} \sum_{n=1}^{N} s_n \qquad (7.12)$$

The centred samples can now be expressed by

$$s' = s - \bar{s} \qquad (7.13)$$

and the unbiased sample covariance matrix is

$$S = \frac{1}{N-1} s' s'^{T} = \frac{1}{N-1}.$$

$$\sum_{n=1}^{N} \begin{bmatrix} s'_n(1)s'_n(1) & s'_n(1)s'_n(2) & \cdots & s'_n(1)s'_n(M) \\ s'_n(2)s'_n(1) & s'_n(2)s'_n(2) & \cdots & s'_n(2)s'_n(M) \\ \cdot & \cdot & \cdots & \cdot \\ \cdot & \cdot & \cdots & \cdot \\ \cdot & \cdot & \cdots & \cdot \\ s'_n(M)s'_n(1) & s'_n(M)s'_n(2) & \cdots & s'_n(M)s'_n(M) \end{bmatrix}$$

where $s'_n(2)$ is the second component of the vector, s'_n. This matrix is symmetric.

7.4   HYPOTHESIS TESTING

In order to determine whether an input image is present when the image is corrupted by noise, the starting point is to posit a hypothesis that the image is not present. This is referred to as the null hypothesis. Testing the null hypothesis involves setting a criterion to decide whether it is true or false. The criterion is called the significance level. Suppose that a filter in an optical correlator has been designed in order to recognise the image. Then the significance level can be the threshold for the height

True Positives	False Positives
False Negatives	True Negatives

Figure 7.4 Confusion matrix

of a given peak in the correlation plane. When the peak is greater than the significance level, then the hypothesis is rejected, and when it is less than, the hypothesis is accepted. If the filter has been correctly designed, the results of the test will correspond to acceptance when the image is not present and rejection when it is present. However, when the noise is pronounced, two kinds of error will arise; false rejection and false acceptance. The former arises when the null hypothesis is rejected but it is true, and the second when it is accepted but is not true. These are called type I and type II errors, respectively. When the signal is a radar return, the type I error is known as a *false alarm*, and the type II error is known as a *miss*. When there is the possibility of more than one type of image in the scene, for example balloons and faces, then the outcomes are as follows. Correct identification of a face is a "True Positive"; balloons detected as faces are "False Positives"; faces in the scene which have been missed because they were confused with balloons are called "False Negatives"; and the correct identification of "not faces" (i.e., balloons) are known as "True Negatives." These can be organized in a confusion matrix (Figure 7.4).

The true and false positives can also be plotted in a receiver operating characteristic (ROC) curve (Figure 7.5). Alternatively, the *misses* can be plotted against the *false alarms* in a detection error trade-off (DET) graph. This typically has the shape of an exponential decay curve, so that a low *false alarm* rate is associated with a high *miss* rate and vice versa. The selection of operating point on the curve can be made via the criterion which is applied. For example, for an image buried in noise, a low threshold will give a large number of *false alarms* and a low number of *misses*. Conversely, a high threshold will give a large number of *misses* and a low number of *false alarms*. Therefore, a given decision boundary will generate data for one point on the curve. As the decision boundary is varied, this generates alternative points on the curve. Returning to the ROC curve, a decision boundary which ranks all round shapes as "hits," will give a point on the top right of the curve; whereas a decision boundary which only admits "full-on" faces, i.e., those with two dark regions at the interpupillary distance of the eyes, will generate a point at the bottom left of the curve. From the point of view of the system designer, the optimal curves are the ones which give the closest approach to the (0,1) point in the ROC curve, and the (0,0) point in the DET curve. In the former case, this gives the highest detection rate (DR) for the lowest false alarm rate (FAR). In the DET curve, the (0,0) point would be the ideal of no *false alarms* and no *misses*. Of the two curves plotted in Figure 7.5, the curve on the left is preferred because it approaches closer to the point (0,1). The two curves can be the result of operating a correlator system with different drive voltages on the filter SLM, for example.

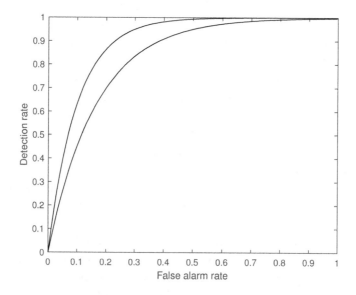

Figure 7.5 Exemplar ROC curves

When there is some variation in the balloons and faces, then a class of each is defined. This is usually based on a limited set of test patterns. In order to correctly attribute new balloons to the class of balloon object which has been defined by the training set, the classifier must have a capacity for generalisation. Patterns within a class have to be recognised as belonging to the same class. This is made more difficult when there are objects such as faces which form a class on their own. The situation can be expressed graphically by forming a feature space, where the axes correspond to common features of the patterns which can be quantified, e.g., size and ellipticity of the face/balloon. The patterns can then be represented by points in this feature space (Figure 7.6).

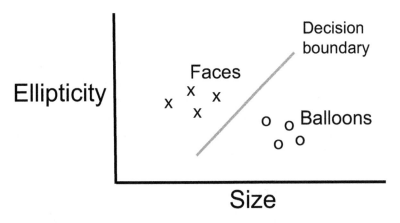

Figure 7.6   Feature space representation of patterns

Multiclass classification is more complex than the simple two-class case described [67]. The two-class case employed a decision boundary which was linear in the feature space selected. This is called a linear discriminant function, and the classifier based upon it is a linear classifier. The parameters of the function can be computed by a machine such as the perceptron (Section 1.1). This is satisfactory for the case presented in Figure 7.6, but will fail on more complex classification problems. The support vector machine (Section 1.1) maps the data into a representation where the classes can be separated. It is used in current machine vision pattern classification.

7.5   FIGURES OF MERIT

In order to enable a comparison between the different filters which can be employed, various performance criteria have been introduced. For some filters the different criteria are antagonistic, in the sense that the decrease of one criterion will lead to the increase of another.

7.5.1 Peak sharpness measures

There are three measures of correlation peak sharpness, the Horner efficiency, the peak-to-correlation energy (PCE), and the peak-to-sidelobe ratio (PSR). The Horner efficiency is a measure which is defined by [105]

$$\eta_H = \frac{C_{(0)}^2}{\int i^2(x,y)dxdy} \tag{7.14}$$

where the numerator is the peak energy (defined in a fixed area of the correlation plane, e.g., 5 x 5 pixel mask), and the denominator is the total energy in the input plane. When the peak energy over a fixed area of the correlation plane is referenced to the total energy in the same plane, then this is known as the PCE [148]. The PSR is the peak energy referenced to a larger area surrounding the peak (e.g., 20 x 20 pixels) from which the area included in the peak energy has been subtracted.

7.5.2 Signal to noise ratio

The signal to noise ratio (SNR) is the ratio of the total intensity in the output plane which is due to the correlation of the input scene and the filter with no noise present, to the total intensity due to the product of the noise and the filter. The product of the noise and the filter is formed as follows. The noise spectrum is squared and multiplied by the square of the filter spectrum. This product is integrated over the frequency domain, and the square root is taken of the resulting integration. The expression for the total intensity in the output plane can be written in terms of the frequency domain representation of the product of the input and the filter function by the use of Parseval's theorem. This theorem states that the total energy of a function is equal to the total energy of its Fourier transform. Therefore, the SNR can be written as a quotient where the numerator is the total energy in the frequency plane of the third term in Equation 6.29. The phase ramp term has been omitted since this provides the displacement and does not contribute to the energy. The denominator is the total integrated energy in the frequency plane of the product of the frequency representations of the noise and the filter function. The noise is the power spectral density of the relevant noise in the system.

7.5.3   Discrimination capability

When the "face" and "not face" (i.e., balloon) targets are well defined, and it is desired to optimise the filter so that the autocorrelation of the face is much greater than the cross-correlation with the balloon, then the discrimination capability figure of merit (DC FOM) should be used. The DC is defined by the difference between the auto and cross-correlations, normalised by the autocorrelation [182].

7.6   FILTER CODING TECHNIQUES

The different types of filter will be defined in the frequency plane representation and identified by an acronym subscript, for example $\hat{F}_{MF}$ for the matched filter. The result of using a frequency domain filter in a hybrid correlator configuration, such as Figure 6.7, is to form the product of the filter with the FT of the input scene, and the FT of this product is the correlator output. As noted in the previous chapter, the correlator output is a superposition of a DC term, and a correlation and convolution of the input scene with the spatial domain filter representation. Therefore, one of the basic aims for a filter design is to maximise the energy density in the correlation term, in particular the peak intensity, because the peak location is commonly detected by a thresholding operation.

The pioneers of optical correlation from within the radar and signal processing communities naturally thought of the matched filter (MF) since this is the one which maximises the SNR for additive "white" noise, which hampers the detection of radar return signals. The MF is the complex conjugate of the frequency spectrum of the template image, $\hat{F}_{MF} = \hat{F}^*(f_x, f_y)$. It is the most important filter in the development of optical correlators, as well as being the basis for the first hand-drawn diffractive element (Chapter 3). Matched filters have a high dynamic range, with high amplitudes at low spatial frequencies, and low amplitudes at high spatial frequencies. In general, the high spatial frequencies contain the signatures of the target image. Therefore, discrimination will be improved if the high spatial frequency content of the filter is enhanced. One method for doing this is the phase-only filter (POF), which is the filter obtained when the amplitude of the MF is set to 1 (Section 7.6.1). Some of the advantages of the POF can be retained if the number of phase levels available for coding is reduced. In the case of a two-level coding, this is called the binary phase-only filter (BPOF) (Section 7.6.2). When the additive noise is not "white" but "coloured", the optimal filter is a Wiener filter which is presented in Section 7.6.3. The MF and Wiener filter demand a complex coding of the filter. If the complex modulation values of the filter are fitted to the operating curve of the SLM, this is known as constrained filtering (Section 7.6.4).

7.6.1 Phase-only filtering

The phase-only filter (POF)is defined by [109]

$$\hat{F}_{POF} = \frac{\hat{F}^*(f_x, f_y)}{|\hat{F}(f_x, f_y)|} \tag{7.15}$$

This filter yields a high Horner efficiency, but the SNR is compromised because it has no noise suppression capability, although some amelioration of this has been devised [146]. The discrimination capability of the POF can be accentuated by attenuating a selected number of spatial frequencies in the filter. This has been employed in the discrimination of an "F" from an "E" in [1]. The high discrimination of this filter is a handicap when there is variability in the template. The design of a POF for this eventuality has also been described [129]. The POF was used in one of the compact correlators listed in Table 6.1 [240]. Although the POF produces a sharper correlation peak for a "hit" than the MF, the filter which optimises the PCE is the inverse filter [148]

$$\hat{F}_{IF} = \frac{1}{\hat{F}(f_x, f_y)} \tag{7.16}$$

A further drawback of the POF is that it is only relevant for images which have a FT with a phase function. In Chapter 1, it was shown that the FT of real symmetric inputs is real (Equation 1.10). If the FT is, furthermore, positive, then the POF would be the null matrix.

7.6.2 Binary phase-only filters

The binary phase-only filter (BPOF) is important due to the availability of fast binary phase SLMs, for example the FLCOS mentioned in Section 2.3. When the polarisation of the light incident on the FLCOS is oriented in the mid-plane between the two orientations of the liquid crystal director, then the two states give $0, \pi$ modulation, respectively. The starting point for the binarisation is the computed MF. The filter can be binarised in a number of ways: for example, binarisation can be imposed on the real part of the MF, or the imaginary part, or the sum of the real and imaginary parts. All the possible binarisations can be incorporated into the phasor description of the filter with the help of a threshold line angle, β,

$$\hat{F}_{BPOF} = 2H[Re(e^{-i\beta}\hat{F}^*(f_x, f_y))] - 1 \qquad (7.17)$$

where H is the Heaviside step function

$$H[n] = \begin{cases} 0, & n < 0, \\ 1, & n \geq 0, \end{cases} \qquad (7.18)$$

When $\beta = 0$, it is the real part of the MF which is binarised; and when $\beta = \pi/2$, it is the imaginary part of the MF which is binarised. A guide to the selection of β is given in [268].

The BPOF was employed in two of the compact correlators listed in Table 6.1 [41, 277]. In a similar manner to the POF, the SNR is compromised in this filter. However, an optimal BPOF, which maximises the SNR, was derived in [75]. The application of a BPOF with a binary amplitude scene input data was explored in [108]. Although the BPOF is particularly well adapted to EASLMs, it has also been written to HRM [39].

7.6.3 Wiener filter

The MF is optimal in the case of additive "white" noise. If the spectral content of the additive noise is not "white" and is known, then the appropriate filter is a Wiener filter. This filter was first described by Norbert Wiener [282],

$$\hat{F}_W = \frac{\hat{F}(f_x, f_y)}{\hat{F}(f_x, f_y) + P_n(f_x, f_y)} \qquad (7.19)$$

where $P_n(f_x, f_y)$ is the spectral distribution of the noise.

7.6.4 Constrained filters

The MF and Wiener filters require a complex coding capability, which was supplied by the HRM used in the VLC systems. No SLMs exist currently where the fully complex coding can be performed by a single pixel. A number of groups around the world are working to develop such SLMs. When this is successfully accomplished, the only constraint on the coding capability of such a device will be the number of levels of modulation, conventionally called the number of grey levels (GLs). In Figure 7.7, the Argand diagrams of four different types of SLM operating characteristic are plotted. The GLs available from the operating characteristic are marked by crosses within and including the unit circle, which represents the maximum modulation capability of the filter. The fully complex capability, Figure 7.7(a), is constrained to 16 GLs, which are selected for approximate uniform coverage. The phase-only capability, Figure 7.7(b), is constrained to 16 GLs evenly spaced on the unit circle. Binary phase-only, Figure 7.7(c), is constrained to 2 GLs on the unit circle. Finally, coupled amplitude/phase, where the 16 GLs are constrained to the measured SLM operating curve, an arbitrary example of which is shown in Figure 7.7(d). This operating curve, where a limited range of phase modulation at low voltages is complemented by amplitude modulation at higher voltages, is commonly found in TN SLMs.

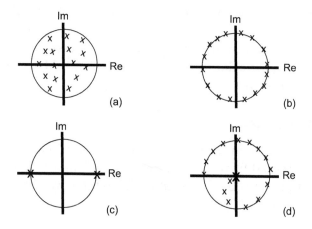

Figure 7.7 Complex plane representations of SLM modulating capability: fully complex (a); phase-only (b); binary phase-only (c); and coupled amplitude/phase (d)

One constraint which has not been represented in Figure 7.7 is binary amplitude. The Argand diagram of this filter consists of two crosses, one at the origin and the other at the intersection of the Real axis and the unit circle. This was employed to code the filters in one of the compact correlators listed in Table 6.1 [55]. This correlator employs a greyscale input coding. The case of binary amplitude input together with binary amplitude filtering has been explored in [147]. In order to constrain a complex amplitude filter function to just two points in the complex plane, the minimum Euclidean distance (MED) principle is used [126]. The Euclidean distance between the point representing the desired complex modulation value and the constrained value is the length of the straight line joining the two values. The distance from each complex modulation value to the constrained values is calculated for all of the values. These distances are then squared and summed to provide what is known as an energy function. The constrained values for every pixel are adjusted until the energy function is a minimum in order to determine the best representation of the filter on the SLM. An additional variable which was used in the calculation of the binary amplitude filter was a rotation angle, ϕ. Each complex modulation value was multiplied by $\exp(i\phi)$ and the best value of ϕ for the minimum energy was selected. The MED principle can be extended to the calculation of the other filters in Figure 7.7.

7.7   FILTERS BASED ON TRAINING IMAGES

When there are a number of samples of the template image, such as a face under varying illumination conditions, the design of the filter for this face should encompass the predominant features of the face. An important question is what constitutes a good set of training images. A variable degree of pre-processing goes into the preparation of the training set, such as co-locating the samples within a fixed grid, thresholding and even edge enhancement. Training images for varying illumination, pose, orientation, and scale should also be treated as different sets, or classes. If the filter can be designed using a good set of training images, and then it correctly classifies an image which is a member of this class but which was not included in the design of the template, this capability is called generalisation. Moreover, filters can be designed for good intra-class recognition and inter-class discrimination. Two sets of techniques have been developed for forming an optimised filter which is representative of a class of objects. The first set uses the techniques of statistical pattern recognition or SPR filters. The second set of techniques grew up and developed within the optical processing community. Since this second set of techniques emphasised the discrimination capability of the filter, they are known as synthetic discriminant function (SDF) filters. A detailed comparison of the two types of filter from a theoretical standpoint was discussed in [258].

7.7.1 SPR filters

In order to design a filter which is representative of a set of training images which are provided, the starting point in SPR is the covariance matrix of these sample images. The covariance matrix is formed in a similar manner to the noise covariance matrix. Each sample image is represented in raster scanned form as an M-dimensional column vector, s, where M is the total number of pixels in the image. The M x M unbiased covariance matrix, S, is then spectrally decomposed into the product of three matrices, $S = E\Lambda E^T$. The eigenvalues along the diagonal are conventionally ordered from maximum to minimum and the corresponding eigenvectors in E and E^T are likewise ordered. Then the predominant features of the template images are contained in the eigenvectors corresponding to the largest eigenvalues. A composite filter based on these eigenvectors will be suitable filter for the recognition of the template in an input scene.

For large M, the spectral decomposition of the covariance matrix can be computationally expensive. When the number of samples of the template image is less than the number of pixels in the image, it is less expensive to perform a SVD of the M x N centred template image matrix, T. If the SVD is $T = Q_1 \Sigma Q_2^T$, then the principal components are $Q_1^T T$, and the procedure for obtaining these is called principal component analysis (PCA). A composite filter can be formed in a similar manner to that formed from the eigenvectors of the covariance matrix, by selecting the k largest eigenvectors. The value of k is determined by either a rapid fall in the value of the eigenvalue between the kth and the (k+1)th, or by the values of those above the (k+1)th being constant, indicating that the noise floor has been reached. PCA is an important technique for separating a sample of images into components which are uncorrelated. The technique can be further refined when it is required to separate the sample into components which are independent. This is known as independent component analysis (ICA), and has been used for face recognition in [7].

Discrimination between two classes of images can be incorporated into this scheme by calculating a covariance matrix which encompasses the training sets of both classes. The eigenvectors can be calculated so that the associated eigenvalues for one class are large when those for the other class are small, and vice versa. The filter corresponding to each eigenvector was correlated with the input image to assess the performance of this technique in [154]. When there are more than two classes, this methodology can be extended [93]. Alternatively, the feature space data can be mapped into a decision space, as in the example of the support vector machine (Section 7.3). The output plane of the correlator is segmented into a number of areas corresponding to the number of classes, so that a spot in one area identifies the class of the object. Since a significant SBWP is required of the filter in this case, CGHs are preferred [94]. Both dimensionality reduction and feature space mapping were used in [44]. The techniques of SPR are valuable when there is significant inter- and intra-class variation. All the above techniques have been applied to the samples of the input object. Two-class discrimination has also been applied in the Fourier domain [71].

7.7.2   SDF filters

The starting point for the development of SDF filters was to make a linear combination of the basis functions of a set of N training images [101]. The M-dimensional column vector representation of each training image T was expanded as the linear combination of a set of orthonormal basis vectors, $T = \Phi B$, where Φ is an M x N matrix of the set of orthonormal basis vectors and B is an N x N matrix of the expansion coefficients. The basis vectors and their coefficients in the linear expansion were calculated by preparing MFs of each of the training images and then measuring the autocorrelation of each with itself and cross-correlation with the other training images. The filter vector, h, can be expressed in terms of Φ, by $h = \Phi c$, where h is the M x 1 filter and c is an N x 1 vector of coefficients. Then, the filter design is the solution of $T^T h = i$ where i is an N x 1 vector of unity elements. This was the equal correlation peak (ECP) SDF. It was pointed out that this expansion can be performed using SVD which allows a dimensionality reduction by selecting only the largest eigenvalues [188].

It was subsequently appreciated that the coefficient matrix, B, and the coefficient vector, c, can be combined into an N x 1 coefficient vector, a, where $a = B^{-1}c$. The SDF filter can be expressed as a linear combination of the training images, $h = Ta$, with the advantage that shift invariance is preserved. These are known as projection SDFs and four have been identified: the ECP for single-class recognition;

the mutual orthogonal filter (MOF) for two-class discrimination; the "nonredundant" for multiclass discrimination with a single filter; and the "K-tuple" for multiclass discrimination with multiple filters [43]. Another ECP SDF was a linear combination of 10 rotated tank images [36]. The DFT of this synthetic image was computed and encoded in filter form using an improved variant of detour phase coding, called the Allebach-Keegan algorithm [8]. The filter was fabricated using an e-beam lithography system. Projection SDFs are robust to "white" noise on the scene input images. They can also be encoded into POFs and BPOFs with increased throughput, sharper correlation peak, and higher SNR [110]. In the case of coloured noise, the minimum variance SDF (MVSDF) is used. This is a linear combination of the training images which have been multiplied by the inverse of the noise covariance matrix [145]. Additive noise and nonoverlapping noise was incorporated into the mean MF for a distortion invariant filter design in [119]. Nonoverlapping noise is everywhere in the input plane except superposed on the target.

Whereas the MF and its POF generate a broad and a sharp correlation peak, respectively, the output plane of a correlator which employs SDFs can have numerous sidelobes, or smaller peaks, which may be mistaken for the main peak. Therefore, an SDF which minimises the energy in the correlation plane, known as the minimum average correlation energy (MACE) was developed [169]. It was subsequently discovered that the noise on the input plane could be also reduced if this filter was designed in the input domain, rather than the frequency domain. This was called the SMACE (space domain MACE) [248]. Some authors prefer to distinguish between filters composed in the space domain and those composed in the frequency domain by reserving SDF for the former. However, the frequency domain MACE filter is here classified under SDF. Some generalization capability was introduced to the MACE filter in the Gaussian-MACE extension [48]. The SNR was improved as well as the correlation plane energy in the MINACE (minimum noise and correlation energy) filter [213]. This filter is also composed in the frequency domain, and was used in vehicle detection in [259]. The final SDF is the kernel SDF [201] which is composed in the space domain using functions of dot products of the training vectors which map the inputs into a convenient decision space.

7.7.3 Trade-off and unconstrained filter design

It is apparent that the crafted filter has to serve a number of roles, such as minimising the correlation plane noise, improving discrimination, and providing generalisation over the set of training images. For each pattern recognition task, the weighting given to each of these roles is different. The trade-off filter accommodates this need by including each factor in the filter calculation with an adjustable parameter [79]. A second type of filter is the unconstrained filter where the hard constraints on the outputs, e.g., the stipulation that these should be 1 at the origin for the ECP SDF, is relaxed. An example of an unconstrained filter design is the maximum average correlation height (MACH) filter which is designed to maximise the correlation peak height for all the training set, whilst simultaneously minimising the output noise and the variance of the training filters [170]. These filters were coded in binary amplitude and used in the GOC (Table 6.1) [55]. It was demonstrated that this filter maximises the PSR in the output plane [168]. The different criteria in the filter design are often traded off and the resulting optimal trade-off MACH (OT-MACH) was used for tracking people in street scenes [19].

7.8 PHASE-ONLY CORRELATION

The previous two sections concentrated on the design of the filter, together with constraints which arise due to the particular filter coding technique which is used. The latter constraints arise from the nature of the SLM used for filtering in the HC architecture. In particular, the phase-only coding of the filter plane has a beneficial aspect on the throughput and discrimination capability of the filter. The reader may now be interested in the effect of the input plane coding on the correlation. Phase-only coding of the input plane produces beneficial results which have been publicised by a number of authors. Phase-only correlation (POC) was the name given to correlators where both the input plane and the filter plane are coded in phase [51]. A large number of simulation studies have been performed [128]. However, the first studies considered the phase coding in the input plane as an undesired side effect of using SLMs with coupled phase/amplitude modulation characteristics [111]. Indications for the adjustments required in the phase-only filter to ameliorate the effects of the coupled phase modulation in the input plane were provided. The first indication that the

SNR can be improved by using a phase-only modulation with a weak associated amplitude modulation was published in simulation studies [107]. A further simulation study was based on the modulation modes of twisted nematic SLMs used as the input device, and the complex conjugate of the input used as the spatial representation of the filter [177]. An important aspect of nonlinear input coding was noted in this paper, namely that additive input noise is converted to nonadditive noise by the device nonlinear coding. The application of POC to fingerprint matching [114] and sub-pixel image registration [252] was demonstrated by digital simulation. The POC technique has been applied for a number of applications using digital processing.

GLOSSARY

Binary phase-only filter (BPOF): A filter which is derived from thresholding the phase-only filter.

Constrained filter: A filter which is coded according to the available complex modultation values of the filter SLM.

Matched filter (MF): The complex conjugate of the frequency representation of the input function.

Phase-only correlation (POC): The correlation between a phase-only representation of the input scene and a phase-only representation of the template.

Phase-only filter (POF): The complex conjugate of the frequency representation of the input function divided by the modulus of the frequency representation.

Receiver operating characteristic (ROC): This is a method for evaluating the performance of a signal detection technique. The ROC graph plots the false-positive rates on the x-axis together with the true-positive rates on the y-axis.

Unconstrained filter design: A filter design procedure where there are no hard constraints on the correlation plane output.

CHAPTER 8

Systems: packaging and applications

CONTENTS

8.1 OPTICAL PACKAGING

The potential of optical interconnects to form large bandwidth highways for data processing was one of the motivations for the large effort on Optical Computing in the 1980s. The advantages are 2D high density wiring which has zero crosstalk even for interpenetrating beams [49]. This potential has still to be fully realised. The following attempts will be described in this section: solid optics, planar (or flat) optics, LEGO®® optics, modular approach, and waferscale assembly.

8.1.1 Systems based on LCLV input transducers

In the 1980s, prior to the establishment of a good source of EASLMs, the input device for the optical correlator was an LCLV. In order for this device to be used in staring mode, the radiant emittance of the input object had to be increased so that its image on the photosensor of the LCLV exceeded the threshold sensitivity of the device. This was achieved by either flashlamp illumination of the input object [62], or an image intensifier attached to the photosensor [262]. In the former system, a photorefractive crystal was used for the Fourier plane filtering, whereas a bleached photographic plate was used in the latter system. These early systems gave a proof of concept and, in the latter case, the system was a precursor to the MOC system [247].

8.1.2 Folded optics

The desire to mount optical correlators inside missile guidance systems was the incentive for stabilising the system against thermal and mechanical perturbations, so that it would withstand extreme environments such as wide temperature ranges and severe vibration levels. The basic VLC was folded into a discoid slab of either fused quartz [9] or Zerodur [166]. Alternatively, solid glass block optics using roof prisms [247], and a similar solid block [137] were the bases of compact correlators. Furthermore, the size was also reduced to allow inclusion on a printed circuit board for embedded processing within a computing system [14]. In addition to the compactness, folded optics on a common substrate gave a robust implementation. However, there remained the problem of accurately aligning the devices on the solid support. With the development of silicon integrated circuit technology for the SLMs, it was appreciated that the two LCOS devices and a portion of the image sensor could be developed upon a common silicon substrate so that the alignment of the folded optics was simplified [193]. A common fabrication technology for these devices ensures that they are in the correct orientation with well-defined spacings between the three arrays. In order to interconnect these three arrays diffractive lenses were employed. The diffractive lenses were also coplanar on a plane adjacent to the SLM/Image sensor plane and the architecture was based on the compact 2f correlator [106]. The necessary path lengths between the components were guaranteed by a mirror/polariser plane at a distance from the diffractive lens plane of approximately f/2, where f is the focal length of the diffractive lens (Figure 8.1). Since the diffractive lens can be engineered to have the precise focal length required for creating a correctly sized Fourier transform of the input on the filter SLM, there is no requirement for creating a synthetic focal length using two refractive lenses, Equation (6.3). An alternative approach is to ensure that the devices are orientationally aligned but their positions may not be precisely known. In this case, the lenses can be constructed holographically to comply with the geometry of the folding [215]. An additional advantage is that the aberrations of the lenses can also be controlled. The capability to represent lens functions on an SLM (Section 4.2) has allowed systems to be constructed where the focal length and aberrations can be corrected under software control. The compactness of the arrangement is limited by the focal length of the SLM lens, which is, in turn, limited by the pixel repeat, due to the requirement to avoid aliasing (Section 4.2). Nevertheless, a folded system based on one LCOS device has been realised [288]. One half of the device displays the input scene multiplexed with a lens, and the second half displays the FT of the template. Due to the phase-only modulation employed in the LCOS device, the multiplexing of a lens function with the input scene is achieved by simply adding the two functions. If the total phase exceeds 2π, then the value is reduced by an integer multiple of 2π. In addition to the lens function added to the input scene, a grating was added to the template half of the device. The grating displaces the correlation signal from the high intensity zero order spot by a distance proportional to the spatial frequency of the grating. The use of one SLM for both the input scene and the template images reduces the SBWP of the system. The adjustability of the arrangement can be maintained with a two SLM system where lens functions are multiplexed on both the scene and the template SLMs [289]. The improved SBWP was utilized for spatially multiplexing both the input scene and the filter function. Spatial multiplexing is practicable when the SBWP of the image is much less than the SBWP of the SLM. In this case, the SBWP of the image was 256 x 256 so that it could be replicated 2 x 2 times across the 792 x 600 pixel device.

The idea of systematically integrating complex optical systems using planar technologies such as reactive ion etching (RIE) lithography or ion-beam lithography was introduced at Bell Technical Labs when there was a concentrated effort on optical computing [115]. This is now known as PIFSO (planar integration of free-space optics). A Fourier transform system has been constructed with PIFSO [183]. The power of DOE design is illustrated by the achromatic nature of the Fourier lens, which is not required in the systems which have been described here.

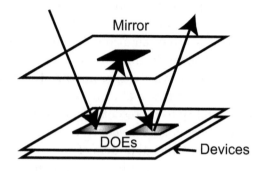

Figure 8.1 Optical correlator folded between two planes with diffractive optical elements (DOEs) and devices combined on one surface with a mirror on the other surface [193]

8.1.3 Modular optics

The optical module is the first level of the integration hierarchy described in [25]. It is the integration of the optoelectronic device array with its optical packaging support structure. An example is the optical engine of a projector. It is constructed as a module in order to fulfil the technical specification of the projector, in particular with regard to illuminance levels. A corresponding optical engine for the correlator would compose the laser diode, collimation optics, beamshaping optics, SLM, beamsplitter, and Fourier lens. The relevant modules for one particular correlator were referred to as sub-assemblies and were designed and toleranced to be drop-in assemblies [247]. For example, the telescope optics for the input LCLV was designed as a unit and subsequently fitted onto the LCLV using machined spacers. The optimum spacer thickness between the telescope and the LCLV was based on the quality of the correlation during system trials. The Fourier lens pair was also a drop-in assembly with a spacer which could be similarly "fine tuned." The folded optical path was composed of prisms which were aligned using an alignment telescope. The critical alignment between the input array and the filter was solved by preparing the HSF within the optical system, developing the emulsion outside the system, and then relocating it within the system using a kinematically registered holder. When two SLMs are used for the input array and the filter, such as the input and frequency plane SLMs in an HC system, the alignment required is more exacting. Translational alignment along the 3 axes and rotational alignment around the same axes (roll, yaw, and pitch) requires a systematic procedure. For example, the roll misalignment around the optical axis can be nulled by comparing the far-field diffraction patterns of the 2 SLMs [189]. Here, the alignment of the SLMs is a precursor to the construction of a high-value solid block module. Failure of one of the SLMs entails the replacement of the complete module. An alternative is to retain the principle advocated in [25], namely, to restrict the module to one device array and to build the alignment into the mating of the modules. In order to accomplish this, the mating surfaces should be either flat or round, and precisely machined with respect to datum surfaces [271].

8.1.4 Optics based on plastic substrates

The main inspiration for choosing to mount optical devices and components on plastic substrates has been their light weight and the availability of low-cost construction technologies such as injection-moulding and rapid prototyping (such as additive manufacturing or 3D printing). The former is based on polymethyl methacrylate (PMMA) or perspex, and the latter is based primarily on acrylonitrile butadiene styrene (ABS). Their coefficients of thermal expansion (CTE) are 6 x 10^{-5} and 9 x 10^{-5} per deg. C, respectively, compared with a CTE of around 8 x 10^{-6} per deg. C for glass. Acrylic substrates are convenient for solid optical modules, and ABS is suitable for a skeleton support structure [195]. The ABS plastic used in LEGO® bricks is injection moulded. LEGO® bricks were used as a support structure for various optical systems [209]. The advantages are that the building blocks are lightweight units with tight geometrical tolerances and strong adhesion, and are easy to click in place. Although integrated fixation features can be machined or built into the plastic assembly, a metal support is preferred for detailed alignment mounts. Therefore, plastic substrates favour the gluing of devices and components. A significant disadvantage of the use of acrylic in solid optical modules is the ease with which the acrylic can be scratched.

8.2 ACTIVE REMOTE SENSING

An early application area for optical correlation was in the analysis of synthetic aperture radar (SAR) data for automatic target recognition (ATR) systems. Ground-based radar is a system which emits an electromagnetic pulse and times the duration until the return signal from the pulse is received. It is used for measuring the distance to aerial features such as aircraft, missiles, clouds, etc. SAR is a coherent imaging technique of greater complexity, involving aircraft or satellite emitters mapping ground-based terrain. Due to the motion of the emitter, the radar is rotated during the flight in order to irradiate the same area of terrain with pulses from different locations on the flight path of the emitter. This provides a higher resolution of the terrain than can be achieved with a stationary emitter. The emitter antenna is effectively lengthened along the flight path, which provides a resolution inversely proportional to the length of the antenna. Due to the collection of radar returns over the length of the antenna, the resulting returns consist of 1D signal slices. These signals must be stored in an onboard data recorder until the availability of a ground station allows transmission. An interesting comparison of analogue (optical) and digital techniques for processing in the ground station was performed over 40 years ago [13]. At the time, analogue processing was established and digital processing was the unknown. The analogue system was known as the tilted plane optical processor. This generated a 2D output image of the terrain, which was limited to a 50 dB dynamic range due to the limitations of recording on photographic film. The contemporary solution to a high dynamic range output device is to use a 2 CCD design where the two arrays have precise pixel registration and each array has its own shutter. This means that one array can be dedicated to the brighter areas of the image and the other can be dedicated to the darker areas. An alternative image-formation processor employed 3 DOEs which performed the required geometric transformation optically in a similar manner to the geometric transformations noted in Chapter 3 [50]. Lower dynamic range targets were identified using a single CCD sensor. An alternative to forming a 2D map of the terrain is to process the 1D signals directly [6, 245]. These optical systems employ 1D SLMs known as acoustooptic scanners. The onboard memory storage and transmission data rates requirements are high, and favour onboard signal processing. Currently, image formation is based on digital computation in earth-based work stations. A significant advantage would result from on-board computational resources. The limited capacity of the on-board data recorder and the data communication network between the craft and ground, coupled with the the limited availability of the ground link, severely restrict the resolution and swath width of the mapping. Compact on-board FPGA-based image formation eliminates the requirement for data compression when the SAR data is transmitted to earth-based computers [249].

8.3 PASSIVE REMOTE SENSING

Passive sensors use solar radiation as the principal source of illumination. Earth-observation low-orbiting imaging systems provide a high-resolution detail of areas of the earth's surface. The terrain is scanned onto single detectors or linear arrays. The first images were collected in the Landsat program in the 1970s using a multi spectral scanner (MSS) built by the Hughes Aircraft Company. Commercial high-resolution imaging has been undertaken by SPOT since the 1980s to the present day. The latest NAOMI camera for SPOT-7 contains CCD arrays built by e2v, one with the format 5000 x 128 pixels operating in frame transfer mode, and four with the format 5000 x 32 pixels operating in time delay integration (TDI) mode. The four were sensitive at four separate spectral regions. The TDI allowed integration of the 5000 pixel line scan along the 32 pixel columns in synchrony with the motion of the satellite, thus enhancing the sensitivity of the line scan image. The camera has a field of view of 60 km and a resolution down to 1.5 m. The spatial greyscale distribution of a remote sensing image is a stationary random process [82]. The GL of each cell is correlated with the neighbouring cells. This is expressed as a correlation function, ϕ, which is the expectation value for the grey levels of cells in a small neighbourhood of M x M cells

$$
\begin{aligned}
\phi(\Delta x, \Delta y) &= \frac{1}{M^2} \sum_{l=1}^{M} \sum_{k=1}^{M} GL(x,y) GL(x + \Delta x, y + \Delta y)) \\
&= \sigma_{GL}^2 exp(-\lambda_h \Delta x - \lambda_v \Delta y) \quad (8.1)
\end{aligned}
$$

where $(\Delta x, \Delta y)$ are the distances between cells in the x- and y- directions, σ_{GL}^2 is the variance of the grey levels (GL), and λ_h, λ_v are the reciprocals of the correlation length in the horizontal and vertical axes of the picture.

The location of target images in remote sensing images is an appropriate application area for the optical correlator. The registration of the target in the scene can be achieved by generating a sequence of sub-images from the scene with a repeat distance which is less than the correlation length [275]. Sub-pixel registration accuracy can be achieved [274]. Pre-processing of the scene image using a thinning algorithm was a pre-requisite for sharp correlation peaks. Each resolution cell embraces a number of point scatterers which give rise to speckle which appears as a "salt and pepper" noise in the image. This can be reduced by median filtering, which replaces every pixel with the median of neighboring pixels. However, a loss of resolution is inevitable. Optical correlation is preferable to filtering because it calculates an intensity based on correlating a complete sub-image area rather than smoothing the image at the individual pixel level; thus reducing noise and maintaining resolution.

Finally, the JTC built by the group at Dresden (Table 6.1) was funded by an ESTEC contract in 2003/2004 which proposed the use of an optical correlator in an active remote sensing satellite. The proposal suggested that the image on the scanning array could be stabilised against satellite vibration using an ancillary optical correlator which monitored the image flow due to the vibration and corrected the scanning system via a tilting mirror.

8.4   DNA SEQUENCING

The genetic make-up of each individual is held in the cell chromosomes. Each chromosome is composed of helical strands of a large DNA (deoxyribonucleic acid) molecule. The building blocks of the DNA are the nucleotides, which give rigidity to the DNA molecule. Each DNA molecule is composed of at least 20,000 nucleotides. The nucleotide molecule is composed of a phosphate group, a sugar, and a nitrogenous base. There are four types of bases,-adenine, cytosine, guanine, and thymine-, which are denoted by the characters 'A', 'C', 'G', and 'T'. In the DNA molecule, they are matched into base pairs, where a base pair (bp) is one of the pairs A-T or C-G. The genetic make-up of a cell, or genome, is coded in the sequence of these bases in the DNA molecule. The determination of this sequence is called DNA sequencing. The detailed sequencing of a genome is a formidable task. The human genome project required an international scientific collaboration working over ten years to complete. In lesser tasks such as DNA fingerprinting, genomic medicine, preconception and prenatal genetic

screening, and targeted drug development, the significant aspect is sequence alignment.

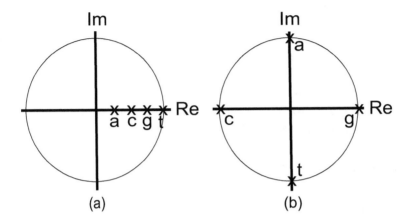

Figure 8.2 Coding of the four bases of DNA for the input plane of an
optical correlator: (a) amplitude coding along the real axis; (b) phase
coding on the unit circle

Sequence alignment is the correlation of a query sequence with a
known database sequence. Query sequences were arranged as 2D images
of varying sizes and correlated with 2D known database sequences in
computer simulation of the VLC [58, 63]. In the latter work, the bases
were coded as grey levels, A as 65, C as 130, G as 195, and T as 255
(Figure 8.2a). The correlation result was compared with the state-of-
the-art basic local alignment search tool (BLAST). BLAST achieves
alignment by first locating the common sequences between the query
and known database sequences. These are small, usually three base, sequences. Alignments based on these common locations are then scored
and the high scoring alignments are then further searched with longer
sequences. This local alignment technique has been preferred to previous global alignment approaches because the latter scale up as the
square of the sequence length and also are sensitive to intrasequence
rearrangements. The significant aspect of the correlation technique for
aligning the sequences was that it was more robust to noise in the
query sequence, which takes the form of individual changes in the bases
at random locations, such as may be induced by genetic mutation.
The robustness of the correlation approach can then be complemented
with fast implementation in optical hardware in order to accommodate the increased computational complexity. Early start-up compa-

nies, such as Optalysis, are endeavouring to apply optical correlation techniques in this direction. Further work on the digital simulation of cross-correlation approaches includes the use of phase-only coding the four bases of both the query and the known sequence [217]. The bases were arranged evenly around the unit circle with the bp arranged in a complementary fashion (Figure 8.2b). The potential for infinities in the phase-only coding with sequences that repeat periodically was solved by zero padding the sequence to a prime number of bases [30]. Finally, an optical correlator based on a 1D acoustooptic scanner was used in a proof of concept system [32].

8.5 SPATIAL FILTERING

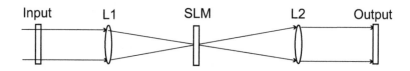

Figure 8.3   4f system layout for spatial filtering using an SLM

The VLC and HC employed spatial filters which facilitated the recognition of an object in a scene. Spatial filtering can be employed to improve the quality of an image as an aspect of scene analysis. In this case, the optical system is less complex. The spatial filter is located at the Fourier transform plane, and the filtered image is reconstituted by a subsequent Fourier transform to the output plane, using a 4f system. The 4f system is known from optical design as a system which gives unity magnification and of which the aberrations can be reduced by using a symmetrical arrangement of lenses. In the imaging application, rays are traced from a point in the input plane to a point in the output plane. If the image in the input plane is conceived as being composed of a spectrum of spatial frequencies, i.e., from a Fourier perspective, then the appropriate ray trace is as shown in Figure 8.3 [243]. Each spatial frequency in the input image diffracts the beam to a separate point on the SLM, where it can be either transmitted or blocked, i.e., the SLM is used in amplitude modulating mode. One important application area is optical microscopy where the optical Fourier transform is already available. The 1953 Nobel Prize lecture of Fritz Zernike was based on his discovery of a spatial filter which advanced the phase and attenuated the intensity of the zero spatial frequency component [290].

Improved contrast of the viewed image resulted from the phase contrast imposed by the spatial filter. When an SLM is used in the spatial filtering plane, more sophisticated spatial filters can be realised [175]. The microscope is preferably equipped with a fibre-coupled laser illumination together with a spinning diffuser for speckle reduction [200]. However, partially coherent illumination can be used in the form of filtered white light which is apertured down so that the spatial coherence is improved sufficiently [176].

Cytopathology involves the visual inspection of cell tissue. It can be performed using spatial filtering in the microscope [199] or on CCD images [95] of tissue. It is an example of a repetitive task where abnormal tissue is a rare occurrence and the false negative rates are, understandably, high. Hence the interest in automating the visual inspection process. An example is cervical cancer screening, where pap-smear slides are prepared from cellular samples obtained from the cervix. Where facilities exist, such slides are sent to a cytology laboratory for staining. This is followed by high-resolution microscopy and automated feature extraction and classification of the cytoplasm and the nucleus. An optical processing approach, consisting of two stages, can be employed to isolate regions of interest (ROIs) for closer (possibly visual) inspection [180]. The first, automatic stage is the recognition of large nuclei in a hybrid correlator, using a filter which is the Fourier transform of a circular dot. Abnormal cells are identified on the basis of their enlarged nuclei. A normal cell's nuclear diameter is between 6 and 8 μm, while an abnormal cell's nuclear diameter ranges between 10 and 22 μm. The circular dot has a diameter which is the lower limit for the diameter of the nucleus in an abnormal cell. The correlation peak width or height can be thresholded to separate the cells with nuclear diameters less than this diameter (normal cells) from those with diameters greater than the filter diameter (abnormal cells). In information processing, simple masks, such as the FT of the circular dot, are called kernels. The convolution of an image with such a kernel is a basic technique in morphological image processing, where 'hits' correspond to those parts of the image that are larger than the dot. Complementary filtering can be performed on the complementary image, where 'misses' correspond to those parts of the complementary image that are smaller than the dot. This involves a complementary kernel to the normal image, and the combination of the two convolutions ('hit' and 'miss') is called the hit/miss transform (HMT). The HMT detects the shape/size of the cell nucleus. If the nucleus is circular with a diameter larger than

10 μm, the cell is identified as suspicious. An optoelectronic implementation of the HMT using two 256 x 256 ferroelectric-liquid-crystal SLMs detects ROIs that can be further processed by ROI classification algorithms. Where advanced facilities for cell screening (staining and high-resolution microscopy) do not exist, for example in the Third World, it has been found that the two-dimensional Fourier transform of the cell is a rich feature space which can be used to differentiate normal and abnormal cells [54].

Two industrial applications in mammography considered Fourier optical systems. Firstly, the correct alignment of ultrasonic and radiological images was suggested as an application for the VLC optical correlator [42]. Secondly, spatial filtering in an optical system was shown to increase lesion detection performance, as evidenced by a reduction in the false negative rate [158]. The latter work was run in a digital simulation of an optical spatial filtering system. The system parameters were based on an upgrade of the sixth-generation module developed at the Lockheed Martin's Advanced Processing Group (Denver, CO). In order to locate ROIs in high-resolution images in near real-time, it was proposed to upgrade the optical system based on 512 x 512 binary ferroelectric SLMs in order to perform 1000 spatial filtering operations per second. Approaches based on purely digital processing can be consulted in order to put this work into context [53, 253].

Another important application of spatial filtering is in texture analysis. It is a task for which the HVS is not particularly well adapted, since there has not been any evolutionary pressure in this direction. It has been shown that the classification of a filtered texture is better than the original unfiltered texture [256]. Both the input pattern and the filter were photographic transparencies which were illuminated by a laser beam. The spatial filter had a passband in the spatial frequency domain. A similar system with an LCLV input has been used for quality control in the paper industry [98]. A statistical method to design the spatial filters for recognition and discrimination between various textures was presented in [170]. A further application area is in defect detection. A Fourier optic system was used to investigate anomalies in a woven pattern [211]. The 4f system illustrated in Figure 8.3 was made more compact by reducing the distances between L1 and the input plane, and L2 and the SLM. Moreover, a beamsplitter was added to the initial Fourier transform so that the FT of the weave could be captured on a CCD. The digital FT was used to detect the positions of three spatial frequencies: the zero order, the warp spatial frequency, and the weft spatial frequency. These were used to calculate four bandpass filters which were displayed on the filter SLM. The bandpass filtering removes the zero order and the warp and weave spatial frequencies so that the output image has less structure than the original image and defects can be found relatively easily by thresholding the digital image. Optical implementation of spatial filtering saves computing time, memory, and storage for high-resolution imaging of wide areas of fabric. A speed on the production line of several metres per second was cited as the justification for optical detection of weave defects in [21]. Both wavelet filtering and Wiener filtering (Section 7.6.3) were employed in the INO optical correlator (Table 6.1), used in spatial filtering mode. In this mode the filter LCTV is operated in amplitude modulation mode. The wavelet selected was the Mexican hat wavelet which is close to the difference of Gaussians (DOG) spatial filter (Figure 8.4), represented by

$$T(f_x) = C(e^{-\frac{f_x^2}{8}} - e^{-\frac{f_x^2}{2}}) \tag{8.2}$$

where $T(f_x)$ is the normalised filter transfer function, C is the normalisation constant, and f_x is the x-component of the spatial frequency. The filter suppresses the low spatial frequencies and acts as a bandpass for the spatial frequencies associated with two types of defect, the pill and floating threads, illustrated in Figure 8.4 as lying between spatial frequencies of 1 and 3.3 lp/mm, and -1 and -3.3 lp/mm. The Wiener filter was configured with the spatial frequency spectrum of the texture as the noise, and the frequency spectrum of the defect as the signal. In both forms of filtering, these defects appear as white blobs on a dark background in the ouput plane. Wholly digital spatial filtering and reconstruction was demonstrated for the density measurement of weave [291]. Here the filtering selected the basic spatial frequencies of the weave so that the pattern repeat can be determined and the density calculated. The lower throughput of this measurement system allowed digital implementation. Finally, spatial filtering can be used in order to improve object recognition in an optical correlator. The AESOP correlator used a DOG spatial filter to bandpass the relevant spatial frequencies which increased the tolerance of the correlator to disparities between the object and reference [232].

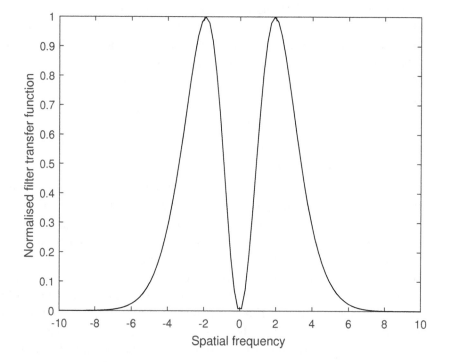

Figure 8.4 Transfer function of difference of Gaussians filter (DOG) versus the x-coordinate of the spatial frequency

8.6   MACHINE VISION

Many inspection tasks both during industrial manufacturing and of finished objects are either hazardous or repetitive. This makes them either unsuitable for human operators, or a tedious human activity which is prone to error by this fact. There has been a strong impetus to replace them by automated inspection, or machine vision (MV). MV is commonly associated with robots, and one of the first systems intended for robotics use, the Autovision 1, dating from 1981, used a 128 x 128 pixel charge injection device (CID) camera. CID cameras were an early competitor to the CCD, from which they differed in the nondestructive nature of their electronic read-out. One of the first application areas for this line of MV systems was in vision-guided welding robots where the human operator has difficulty in viewing the welding arc due to the high intensity of the arc itself. Prior to this, General Motors Corporation investigated the application of a VLC to the inspection of machined parts [87]. The input transducer in this system was an LCLV and the matched filter was recorded on an Agfa 10E75 photographic plate. The plate was held in a liquid gate which is a cell with optically flat windows which is filled with o-xylene. The liquid gate improves the optical quality of the developed photographic plate, in particular with respect to the possible reticulation of the emulsion during development. Its primary benefit is the reduction of spatial noise. Liquid gates are marketed by companies which manufacture film scanners and film recorders. They also allow precise replacement of the developed hologram. The beam intensity ratio (Section 6.4.2) was selected to be 1:1 at the spatial frequency of 3 cycles/mm for the particular object selected. The spatial invariance of the correlator was demonstrated, and the accuracy of the object location was ± 0.4 mm.

A further aspect of MV is range sensing for which optical techniques were explored by the International Business Machines (IBM) research laboratory [246]. One technique which may be of interest to readers utilises the systematic variation in the Fresnel diffraction pattern of a grating as a function of propagation distance [56]. A more recent technique for range sensing is to use lidar (LIght Detection And Ranging), also known as ladar (LAser Detection And Ranging). Ladar was used by the Air Force Research Laboratory to construct a variety of 2D images of a tank (varying the distance and azimuth) at around 200 m distance [38]. The resulting images were preprocessed using noise removal and edge enhancement, and, from these images, composite POFs were constructed. When these composite filters, complemented with a zero order amplitude block, were used in an HC, the tanks were successfully recognised over the range of azimuths used in creating the composite filter. The design of POFs with an amplitude block on the low spatial frequencies is covered in [80] and heuristics for the size of the block are given in [127]. Currently, lidar is used by Google Earth for creating 3D models, and also in self-driving cars for collision avoidance with respect to other vehicles, pedestrians, or unexpected objects. It is also used in the emerging field of robot localization, where cross-correlation must be used in order to match the lidar scan with a known map or reference scan. For mobile applications, there has been an emphasis on simplifying the computational load of the cross-correlation calculation.

The use of an optical correlator inside a moving vehicle was mentioned in Section 6.7. More recent work has addressed the problem of road mark detection and recognition [196]. In a simulation of a VLC employing a POF, the authors demonstrate that the arrows and bicycle markings on street roads can be detected and recognized. Since the correlation plane of a POF correlator is noisy, a good FOM to use is the PCE. The identification of the particular mark was confirmed using a HOG feature detector followed by a SVM classifier. Finally, template matching was found to provide the best score for the recognition of characters in number plates after they had been segmented [257].

An elegant example of the power of Fourier optics is the discovery of defects in a repetitive pattern. The repetitive pattern in this case was a thin metal shadow mask which was an essential component of the cathode ray tube television. The dimension of the shadow mask examined was a 14 inch diagonal with 80,000 holes each of 200 μm diameter [135]. When illuminated by a collimated laser beam, the Fourier transform of this shadow mask can be spatially filtered by a piece of aluminium foil with a hole. When the foil is placed in the Fourier plane so that the hole allows through the zero order and low spatial frequencies, it acts as a low-pass filter. Provided the foil blocks the spatial frequencies corresponding to the regular hole pattern, the output image displays the defects which can be readily observed because the regular pattern of the holes has been removed, in a similar manner to the removal of the regular weave pattern. The spatial frequency spectrum can also be used for low cost, quality control metrology. The thickness, parallelism, and surface irregularity of a silicon wafer bar was measured by a camera in the Fourier plane [251]. An MV task was selected to illustrate the application of the Brite correlator listed in Table 6.1. The optical memory was ideal for the storage of the templates of a machine part at all possible orientations. These could be read out at high speed in order to establish the orientation of the part in a scene.

Spectral filtering of the scene can be advantageous in product inspection. Boulder Nonlinear Systems worked on an HC for the inspection of produce such as apples [16]. An electronically switchable liquid-crystal tunable filter, placed in front of a CCD camera, is employed to select the frequency of imagery that is brought into the correlator. The filter could be switched between different wavelengths in the visible and near-infrared ranges within 25 μs. Certain wavelengths enhance defects in the produce. Two examples given in the paper were scabs on a red apple which are enhanced with red filtered light and sub-surface bruising which is detected in the infrared. Subsequent detection of the defect in the correlator signals removal of that apple from the line of produce which is being inspected. The paper also discusses the two types of LCOS which were used in the correlator. The configurations of the liquid crystal layer in these two types of LCOS were PAN and FLC. Moreover, the paper catalogues how the modulation of amplitude and phase generated by each of these configurations can be varied according to the orientation of retardation plates placed in front of the LCOS device in the paths of the incident and reflected beams.

8.7 BIOMETRICS

The market for biometric systems is expected to exceed $20 billion by 2020. Biometric identification is used to confirm the identity of a person for security purposes. Three prominent types of identification systems are based on the uniqueness of the fingerprint, the iris, or the face. Fingerprints are probably the most widely used. They have been used for identification purposes for over 100 years. A database of fingerprints, FVC2004, is included in the handbook [172]. For a limited access system, such as the workers in a laboratory, a straightforward correlation of the print with a limited database can be successful. I personally verified the accuracy of a system employed to identify the fingerprints of workers in the Hamamatsu research company. This system was based on a joint transform correlator (JTC) [140]. For more demanding recognition tasks, a digital simulation of the phase-only correlation (POC) technique (Section 7.8) has been used [114]. The POC is particularly useful in those cases where the captured fingerprint is of poor quality [113]. This was successfully tested on a larger factory of 700 employees.

The POC digital simulation has been used to coarsely assign two fingerprints, prior to detailed matching in a convolutional neural network in [208]. An alternative to fingerprints is to use the finger vein. The network pattern of the finger vein is distinct for each individual and is internal. An improved finger-vein identification algorithm based on template matching was proposed and confirmed experimentally [237]. A digital simulation of POC has also been used for iris recognition with a view to a digital signal processing implementation [184].

Face recognition software is commonly used on digital cameras to locate faces in scenes. Face authentification by digital approaches is an emerging biometric, which has been accelerated by a recent trend in 3-D face mapping. Distinctive features such as the eye socket contour, and the chin and nose shape are not subject to variation with illumination conditions. They can also be measured under a variety of angles of the face, rather than a straight-ahead shot. The face identification system on the Apple iPhone projects more than 30,000 infrared dots onto the user's face. The projected grid method yields an improved recognition accuracy of 1 in a million false positives, where the wrong person would be allowed to unlock the phone. Qualcomm has a computer vision kit that is capable of active depth sensing, using an infrared illuminator, infra-red camera and a 16-megapixel RGB camera. The illuminator fires a light that creates a dot pattern (using a filter), and the infra-red camera searches for and reads the pattern. By calculating how the dots warp over a subject and the distance between points, the system can estimate the 3D profile of faces. Identification of faces from a limited database has been claimed by NEC, based on their NeoFace software. Interestingly, a multichannel approach to face localization based on correlating 10 landmark features across the face was outlined in [86]. The approach was extended to car detection based on the MIT Streetscene database, and to pedestrian detection based on the Daimler pedestrian dataset. The interest of the approach is, firstly, that the HOG low level descriptors (Chapter 1) are used to pre-process the images for correlation, and, secondly, the memory and computational cost of performing correlation in the spatial frequency domain scales favourably compared with alternative approaches when the number of training images increases.

Face identification using the 2D correlation approach follows the capture by a standard camera and segmentation and excision of the faces from the scene. The identification task, in this case, is compounded by varying pose, illumination, and expression. In order to address this variation, the PIE (pose, illumination, and expression) database has been compiled by the Robotics Institute at Carnegie Mellon University [238]. Prior to the establishment of such databases, the training sets were established in an ad hoc manner. Turk and Pentland used PCA of four centred, segmented faces of each of ten individuals, and the resulting linear combinations of the most important eigenvectors were called Eigenfaces [261]. A novel technique developed by Vijaya Kumar was to perform the PCA on the phase-only Fourier transforms of the training images. This was called Eigenphases, and the results compared favourably with the eigenface technique [229]. POFs of the independent components of the face images of the training set were used to compare the correlation results of the known training set images with the results of an unknown face in a simulation experiment [7]. A trade-off filter design was tested with good results on the DC performance metric in [203]. An alternative to the PCA is to correct all face input images by an affine transform in a pre-processing step [278]. The four reference points of the face are the eyes and nostrils. These allow the face to be centred, rotated and scaled. The pre-processing is completed by edge enhancement and binarisation. Although the total pre-processing time is 200 ms, this delay is only incurred for the input image which is placed in the filter plane. All the images stored in memory which have been pre-processed can be input to the correlator at relatively high speed and compared with the unknown face until a match is found. Pre-processing is used in the simulations performed in [186], where varying illumination conditions can be surmounted by using DOG and Local Binary Patterns preprocessing. The latter replaces the intensity value of a given pixel by an 8-bit binary word, where the 8 bits describe whether the eight neighbouring pixels have a higher intensity (1) or a lower intensity (0) than the given pixel. The PIE, AMP, and FIA databases, from Carnegie Mellon University, are three of many face image databases; others include Faces in the Wild, the Yale Face Database, Pointing Head Pose (PHPID), the Georgia Institute of Technology Face Database, and FERET. Some are freely available and others are available on request. They allow the researcher to test correlator configurations by simulation, and develop appropriate encoding techniques for both the input plane and the filter plane.

Biometric Encryption^{TM} links a cryptographic key with a biometric such as the fingerprint or a face (Chapter 22 of [190]). The cryptographic key is necessary for access to some form of digital asset. The output of the optical correlator, in this instance, is not a correlation peak but the cryptographic key. In order to increase the security of the system, an extra feature is incorporated. The extra security feature is the bonding of a phase mask to the primary identification amplitude pattern such as the fingerprint, picture of a face, or signature [118]. The phase mask and the primary pattern are separately identifiable in an optical correlator.

8.8 TARGET TRACKING

In order to track an object in a scene, extra functionality of the recognition system is required. It should be able to locate the object in a reasonable time so that it does not move a large distance between the frames where it can be located by the correlator. Additionally, the object may become occluded or pass similar objects, so that the tracking should be robust against this influence. Furthermore, it may be required to track multiple objects. Commonly, target tracking is performed using infrared cameras, known as Forward Looking Infra Red (FLIR), with lower resolution and frame speed than the image sensors discussed in Chapter 4. Due to the target motion, the filter must accommodate the different orientations and ranges of the object. This is known as an adaptive template. The template can either be updated to account for changes or the foreseeable changes can be incorporated into the filter function. An example of the former is the template matching approach adopted in [29]. When the correlation output drops, the template is optimised to recover the correlation peak amplitude. This procedure was adopted to complement the use of a JTC in simulation [4]. An example of the latter was the tracking of moving tanks using filters based on the MACH filter (Chapter 7) [5]. The filters were trained on the expected size and orientation variations of the object. An approach where the object detection and tracking are integrated by applying probabilistic models to the information in correlation outputs before thresholding was presented [131]. A similar model was the basis for an activity detection correlator [171]. More recent approaches have moved away from the use of optical correlation because of the low complexity of the infrared image of the target and the tracking itself. Once the object location has been determined and the next location of the object has been estimated, a ROI is created which is small enough for sum tables to be used for localisation. This is the case for the intensity variation function, which was successfully used for pedestrian tracking [149]. The location of people in a crowd has been addressed using a MACH filter which is based on the silhouette of the head and shoulders in all orientations and aspects [214].

Unmanned autonomous vehicles (UAVs) operate at low altitudes, for example down to 10 m above the terrain, and are equipped with a 2D visible wavelength image sensor complemented by a fish-eye lens with a wide field of view. In order to navigate the UAV, one option is to generate a digital elevation map (DEM), which is a running 2.5D model of the terrain [255]. An efficient means for generating the DEM using a JTC is also described. It consists of dividing the scene into a large number of sub-images, which can be tracked from frame-to-frame in order to generate an optical flow field. The optimal size of the sub-images was 24 x 24 pixels and the JTC was capable of 50,000 correlations per second [254].

8.9   SATELLITE NAVIGATION

The potential of optical processing for lightening the satellite payload was recognized by early workers in this field. The optical correlator is particularly suitable for a single-vision function, such as the Soyuz docking on the International Space Station [17]. The programmable filter in the optical correlator was particularly useful for changing the scale of the filter as the Space Station was approached. A second application researched in the same paper was the identification of a landing site using a POF. Finally, star tracking was demonstrated on images from the Mount Wilson Observatory. Onboard star tracking is important for attitude determination. Previously, JPL had worked on a system for attitude determination using fixed templates in a multichannel correlator [230]. The contemporary equivalent of these early experiments is a system for aligning the X-ray telescope in a balloon payload [161]. The star camera is aligned with an on-board alignment system using a JTC to align LED arrays on the two systems. The JTC is implemented digitally using fastest Fourier transform in the West (FFTW) software implemented on an NVidia GPU. The implementation was of sufficient precision to allow good alignment, and presumably benefitted from the sparse nature of the matrices involved in the FTs. Finally, tracking across an observed planet surface using an optically implemented JTC with fast update rate was demonstrated [116].

Bibliography

[1] E. Ahouzi, J. Campos, and M. J. Yzuel. Phase-only filter with improved discrimination. *Opt. Lett.*, 19(17):1340–1342, Sep 1994.

[2] G. B. Airy. On the Diffraction of an Object-glass with Circular Aperture. *Transactions of the Cambridge Philosophical Society*, 5:283–291, 1835.

[3] B. Akin, P. A. Milder, F. Franchetti, and J. C. Hoe. Memory bandwidth efficient two-dimensional Fast Fourier transform algorithm and implementation for large problem sizes. *IEEE 20th Annual International Symposium on Field-Programmable Custom Computing Machines (FCCM)*, pages 188–191, April 2012.

[4] M. S. Alam and A. Bal. Improved multiple target tracking via global motion compensation and optoelectronic correlation. *IEEE Transactions on Industrial Electronics*, 54(1):522–529, Feb 2007.

[5] Mohammad S. Alam and Sharif M. A. Bhuiyan. Trends in correlation-based pattern recognition and tracking in forward-looking infrared imagery. *Sensors*, 14(8):13437–13475, 2014.

[6] Carl C. Aleksoff, Ivan Cindrich, Nikola S. Subotic, T. Blessing, and L. M. Przebienda. Optical processor for SAR image formation: system development. *Proc. SPIE*, 2236, 1994.

[7] A. Alfalou and C. Brosseau. Robust and discriminating method for face recognition based on correlation technique and independent component analysis model. *Opt. Lett.*, 36(5):645–647, Mar 2011.

[8] J. P. Allebach. Representation-related errors in binary digital holograms: a unified analysis. *Appl. Opt.*, 20(2):290–299, Jan 1981.

[9] R.H. Anderson. Optical correlator with symmetric reflective optics, September 15 1992. US Patent 5,148,496.

[10] David Armitage, Ian Underwood, and Shin-Tson Wu. *Introduction to Microdisplays*. John Wiley & Sons, Ltd, 2007.

[11] Steven M. Arnold. E-beam Written Computer Generated Holograms. Technical Report AFOSR-TR-83-0850, Honeywell Corporate Technology Center, 10701 Lyndale Avenue South Bloomington, Minnesota 55420, August 1983.

[12] P. Aubourg, J. P. Huignard, M. Hareng, and R. A. Mullen. Liquid crystal light valve using bulk monocrystalline $Bi_{12}SiO_{20}$ as the photoconductive material. *Appl. Opt.*, 21(20):3706–3712, Oct 1982.

[13] Dale A. Ausherman. Digital vs. optical techniques in synthetic aperture radar data processing. *Proc. SPIE*, 0119:238–257, 1977.

[14] H.R. Bagley, J.A. Sloan, and D.W. Small. Vander Lugt optical correlator on a printed circuit board, August 19 1997. US Patent 5,659,637.

[15] D. H. Ballard, G. E. Hinton, and T. J. Sejnowski. Parallel visual computation. *Nature*, 306:21–26, 1983.

[16] K. A. Bauchert, S. A. Serati, G. D. Sharp, and D. J. McKnight. Complex phase/amplitude spatial light modulator advances and use in a multispectral optical correlator. 3073:170–177, March 1997.

[17] A. Bergeron, P. Bourqui, and B. Harnisch. Lightweight compact optical correlator for spacecraft docking. *Proc. SPIE*, 6739, October 2007.

[18] Alain Bergeron. Optical correlator for industrial applications, quality control and target tracking. *Sensor Review*, 20(4):316–321, 2000.

[19] P. M. Birch, W. Hassan, R. C. D. Young, and C. R. Chatwin. Human tracking with multiple parallel metrics. In *International Conference on Imaging for Crime Detection and Prevention 2013 (ICDP 2013)*. Kingston University, December 2013.

[20] Philip M. Birch, Frederic Claret-Tournier, David Budgett, Rupert Young, and Christopher Chatwin. Optical and electronic design of a hybrid digital-optical correlator system. *Optical Engineering*, 41(1):32–40, 2002.

[21] Nathalie Blanchard, Donald Prevost, and Yunlong Sheng. Optical correlator for textile web defect detection. *Proc. SPIE*, 4043:351–359, 2000.

[22] B. A. F. Blandford. A New Lens System for use in Optical Data-Processing. In J. H. Dickson, editor, *Optical Instruments and Techniques*, pages 435–443, 1970.

[23] W. P. Bleha and P. F. Robusto. Optical-to-optical image conversion with the liquid crystal light valve. *Proc. SPIE*, 0317:179–184, 1982.

[24] Brett D. Bock, Thomas A. Crow, and Michael K. Giles. Design considerations for miniature optical correlation systems that use pixelated input and filter transducers. *Proc. SPIE*, 1347:297–309, 1990.

[25] G.C. Boisset. *Optomechanics and optical packaging for free-space optical interconnects.* PhD Thesis, McGill University, 1997.

[26] M. Born and E. Wolf. *Principles of Optics.* Pergamon Press, 1983, 6th edition, 1983.

[27] Robert W. Brandstetter and Nils J. Fonneland. Photopolymer elements for an optical correlator system. *Proc. SPIE*, 1559:308–320, 1991.

[28] Peter Brick and Christopher Wiesmann. Optimization of LED-based non-imaging optics with orthogonal polynomial shapes. *Proc. SPIE*, 8485, 2012.

[29] K. Briechle and U. D. Hanebeck. Template matching using fast normalized cross correlation. In *Proceedings of SPIE: Optical Pattern Recognition XII*, volume 4387, pages 95–102, March 2001.

[30] A.K. Brodzik. System, method and computer program product for DNA sequence alignment using symmetric phase only matched filters, August 7, 2012. US Patent 8,239,140.

[31] N. Brousseau and H. H. Arsenault. Emulsion thickness and space variance: combined effects in Vander Lugt optical correlators. *Appl. Opt.*, 14(7):1679–1682, Jul 1975.

[32] N. Brousseau, R. Brousseau, J. W. A. Salt, L. Gutz, and M. D. B. Tucker. Analysis of DNA sequences by an optical time-integrating correlator. *Appl. Opt.*, 31(23):4802–4815, Aug 1992.

[33] B. R. Brown and A. W. Lohmann. Complex spatial filtering with binary masks. *Appl. Opt.*, 5(6):967–969, Jun 1966.

[34] Olof Bryngdahl. Geometrical transformations in optics. *J. Opt. Soc. Am.*, 64(8):1092–1099, Aug 1974.

[35] J. D. Burcham and J. E. Vachon. Cueing, tracking, and identification in a maritime environment using the ULTOR optical processor. *Proc. SPIE*, 5780:164–172, May 2005.

[36] Steve Butler and Jim Riggins. Simulation of synthetic discriminant function: Optical implementation. *Optical Engineering*, 23:721–726, 1984.

[37] H. J. Butterweck. General theory of linear, coherent optical data-processing systems. *J. Opt. Soc. Am.*, 67(1):60–70, Jan 1977.

[38] David Calloway and Dennis H. Goldstein. Ladar image recognition using synthetically derived discrete phase-amplitude filters in an optical correlator. *Proc. SPIE*, 4734:90–101, 2002.

[39] J. Campos, B. Janowska-Dmoch, K. Styczynski, K. Chalasinska-Macukow, F. Turon, and M. J. Yzuel. Computer-generated binary phase-only filters with enhanced light efficiency recorded in silver halide sensitized gelatin. *Pure and Applied Optics: Journal of the European Optical Society Part A*, 2(6):595, 1993.

[40] E. Carcole, J. Campos, and S. Bosch. Diffraction theory of Fresnel lenses encoded in low-resolution devices. *Appl. Opt.*, 33(2):162–174, Jan 1994.

[41] David T. Carrott, Gary L. Mallaley, Robert B. Dydyk, and Stuart A. Mills. Third-generation miniature ruggedized optical correlator (MROC) module. *Proc. SPIE*, 3386:38–44, 1998.

[42] D.T. Carrott and T.M. Burke. Optical correlator based automated pathologic region of interest selector for integrated 3D ultrasound and digital mammography, May 28 2002. US Patent 6,396,940.

[43] D. Casasent, W. Rozzi, and D. Fetterly. Projection synthetic discriminant function performance. *Optical Engineering*, 23(6):716–720, 1984.

[44] D. Casasent and V. Sharma. Feature extractors for distortion-invariant robot vision. *Optical Engineering*, 23(5):492–498, 1984.

[45] David Casasent. Performance evaluation of spatial light modulators. *Appl. Opt.*, 18(14):2445–2453, Jul 1979.

[46] David Casasent and Alan Furman. Sources of correlation degradation. *Appl. Opt.*, 16(6):1652–1661, Jun 1977.

[47] David Casasent and R. L. Herold. Novel hybrid optical correlator: Theory and optical simulation. *Appl. Opt.*, 14(2):369–377, Feb 1975.

[48] David Casasent, Gopalan Ravichandran, and Srinivas Bollapragada. Gaussian–minimum average correlation energy filters. *Appl. Opt.*, 30(35):5176–5181, Dec 1991.

[49] H. J. (Henry John) Caulfield and Christopher Tocci. *Optical interconnection: foundations and applications*. Boston: Artech House, 1994.

[50] Jack N. Cederquist, Michael T. Eismann, and Anthony M. Tai. Holographic polar formatting and realtime optical processing of synthetic aperture radar data. *Appl. Opt.*, 28(19):4182–4189, Oct 1989.

[51] K Chalasinska-Macukow and E Baranska. Discrimination of characters using phase information only. *Journal of Optics*, 21(6):261–266, 1990.

[52] K. Chalasinska-Macukow and T. Szoplik. Reconstruction of a fraunhofer hologram recorded near the focal plane of a transforming lens: experiment and results. *Appl. Opt.*, 20(8):1471–1476, Apr 1981.

[53] Heang Ping Chan, Jun Wei, Yiheng Zhang, Mark A. Helvie, Richard H. Moore, Berkman Sahiner, Lubomir Hadjiiski, and Daniel B. Kopans. Computer-aided detection of masses in digital tomosynthesis mammography: Comparison of three approaches. *Medical Physics*, 35(9):4087–4095, 2008.

[54] Thanatip Chankong, Nipon Theera-Umpon, and Sansanee Auephanwiriyakul. *Cervical Cell Classification using Fourier Transform*, pages 476–480. Springer Berlin Heidelberg, Berlin, Heidelberg, 2009.

[55] T.H. Chao, H. Zhou, and G. Reyes. Spacecraft navigation using a grayscale optical correlator. In *Proceeding of the 6th International Symposium on Artificial Intelligence and Robotics and Automation in Space: i-SAIRAS 2001*, May 2001.

[56] P. Chavel and T. C. Strand. Range measurement using Talbot diffraction imaging of gratings. *Appl. Opt.*, 23(6):862–871, Mar 1984.

[57] Y. H. Chen, T. Krishna, J. Emer, and V. Sze. Paper 14.5: Eyeriss: An energy-efficient reconfigurable accelerator for deep convolutional neural networks. *IEEE International Solid-State Circuits Conference (ISSCC)*, pages 262–263, Jan 2016.

[58] William A. Christens-Barry, James F. Hawk, and James C. Martin. Vander Lugt correlation of DNA sequence data. *Proc. SPIE*, 1347:221–230, 1990.

[59] Jerome Colin, Nicolas Landru, Vincent Laude, Sebastien Breugnot, Henri Rajbenbach, and Jean-Pierre Huignard. High-speed photorefractive joint transform correlator using nonlinear filters. *Journal of Optics A: Pure and Applied Optics*, 1(2):283–285, 1999.

[60] N. Collings, R. C. Chittick, I. R. Cooper, and P. Waite. A hybrid optical/electronic image correlator. In *Laser/Optoelektronik in der Technik / Laser/Optoelectronics in Engineering: Vorträge des 8. Internationalen Kongresses / Proceedings of the 8th International Congress Laser 87 Optoelektronik*, pages 204–208. Springer Berlin Heidelberg, Berlin, Heidelberg, 1987.

[61] N. Collings, W. A. Crossland, P. J. Ayliffe, D. G. Vass, and I. Underwood. Evolutionary development of advanced liquid crystal spatial light modulators. *Appl. Opt.*, 28(22):4740–4747, Nov 1989.

[62] Neil Collings. *Optical pattern recognition using holographic techniques.* Wokingham, England: Addison-Wesley Pub. Co, 1988.

[63] Millaray Curilem Saldjas, Felipe Villarroel Sassarini, Carlos Munoz Poblete, Asticio Vargas Vasquez, and Ivan Maureira Butler. Image Correlation Method for DNA Sequence Alignment. *PLoS ONE*, 7(6):e39221+, June 2012.

[64] N. Dalal and B. Triggs. Histograms of oriented gradients for human detection. *IEEE Computer Society Conference on Computer Vision and Pattern Recognition*, 1:886–893, June 2005.

[65] Jeffrey A. Davis, Mark A. Waring, Glenn W. Bach, Roger A. Lilly, and Don M. Cottrell. Compact optical correlator design. *Appl. Opt.*, 28(1):10–11, Jan 1989.

[66] N. Douklias and J. Shamir. Relation between Object Position and Autocorrelation Spots in the Vander Lugt Filtering Process. 2: Influence of the Volume Nature of the Photographic Emulsion. *Appl. Opt.*, 12(2):364–367, Feb 1973.

[67] Richard O. Duda, Peter E. Hart, and David G. Stork. *Pattern Classification (2nd Edition)*. Wiley-Interscience, 2000.

[68] M. Duelli, A. R. Pourzand, N. Collings, and R. Dandliker. Pure phase correlator with photorefractive filter memory. *Opt. Lett.*, 22(2):87–89, Jan 1997.

[69] M. Duelli, A. R. Pourzand, N. Collings, and R. Dandliker. Holographic memory with correlator based readout. *IEEE Journal of Selected Topics in Quantum Electronics*, 4(5):849–855, Sep 1998.

[70] J. G. Duthie, C. R. Christensen, R. D. McKenzie, Jr., and J. Upatnieks. Real-time optical correlation with solid-state sources. In W. T. Rhodes, editor, *1980 International Optical Computing Conference I*, volume 231 of *Proc. SPIE*, pages 281–290. January 1980.

[71] J. Duvernoy. Optical pattern recognition and clustering: Karhunen-Loeve analysis. *Appl. Opt.*, 15(6):1584–1590, Jun 1976.

[72] Uzi Efron, W. R. Byles, Norman W. Goodwin, Richard A. Forber, Keyvan Sayyah, Chiung Sheng Wu, and Murray S. Welkowsky. Charge-coupled-device-addressed liquid-crystal light valve: an update. *Proc. SPIE*, 1455:237–247, 1991.

[73] European Machine Vision Association. EMVA Standard 1288. http://www.emva.org/wp-content/uploads/EMVA1288-3.0.pdf. Accessed: 3 November 2016.

[74] Teresa Ewing, Steven A. Serati, and Kipp Bauchert. Optical correlator using four kilohertz analog spatial light modulators. *Proc. SPIE*, 5437:123–133, 2004.

[75] Michael W. Farn and Joseph W. Goodman. Optimal binary phase-only matched filters. *Appl. Opt.*, 27(21):4431–4437, Nov 1988.

[76] Michael W. Farn, Margaret B. Stern, and Wilfrid Veldkamp. The making of binary optics. *Opt. Photon. News*, 2(5):20–22, May 1991.

[77] J. R. Fienup. Phase retrieval algorithms: a comparison. *Appl. Opt.*, 21(15):2758–2769, Aug 1982.

[78] James R. Fienup. Phase retrieval algorithms: a personal tour. *Appl. Opt.*, 52(1):45–56, Jan 2013.

[79] J. Figue and Ph. Refregier. Optimality of trade-off filters. *Appl. Opt.*, 32(11):1933–1935, Apr 1993.

[80] David L. Flannery, John S. Loomis, and Mary E. Milkovich. Transform-ratio ternary phase-amplitude filter formulation for improved correlation discrimination. *Appl. Opt.*, 27(19):4079–4083, Oct 1988.

[81] Eric R Fossum et al. CMOS image sensors: electronic camera-on-a-chip. *IEEE transactions on electron devices*, 44(10):1689–1698, 1997.

[82] L. E. Franks. A model for the random video process. *The Bell System Technical Journal*, 45(4):609–630, April 1966.

[83] Yann Frauel and Bahram Javidi. Digital three-dimensional image correlation by use of computer-reconstructed integral imaging. *Appl. Opt.*, 41(26):5488–5496, Sep 2002.

[84] Edward Friedman and John Lester Miller. *Photonics rules of thumb: optics, electro-optics, fiber optics, and lasers.* McGraw-Hill, 2004.

[85] D. Gabor. A new microscopic principle. *Nature*, 161(4098):777–778, May 1948.

[86] H. K. Galoogahi, T. Sim, and S. Lucey. Multi-channel correlation filters. In *2013 IEEE International Conference on Computer Vision*, pages 3072–3079, Dec 2013.

[87] Aaron D. Gara. Real-time optical correlation of 3-D scenes. *Appl. Opt.*, 16(1):149–153, Jan 1977.

[88] A Georgiou, J Christmas, N Collings, J Moore, and W A Crossland. Aspects of hologram calculation for video frames. *Journal of Optics A: Pure and Applied Optics*, 10(3):035302–035310, 2008.

[89] R. W. Gerchberg and W. Owen Saxton. A practical algorithm for the determination of the phase from image and diffraction plane pictures. *Optik*, 35:237–246, 1972.

[90] Joseph Goodman. *Introduction to Fourier Optics.* Roberts and Company Publishers, 3rd edition, 2005.

[91] J. M. Gordon and Ari Rabl. Reflectors for uniform far-field irradiance: fundamental limits and example of an axisymmetric solution. *Appl. Opt.*, 37(1):44–47, Jan 1998.

[92] Jan Grinberg, Alex Jacobson, William Bleha, Leroy Miller, Lewis Fraas, Donald Boswell, and Gary Myer. A new real-time noncoherent to coherent light image converter the hybrid field effect liquid crystal light valve. *Optical Engineering*, 14(3):217–225, 1975.

[93] Z.-H. Gu and S. H. Lee. Optical Implementation of the Hotelling Trace Criterion for Image Classification. *Optical Engineering*, 23:727, December 1984.

[94] Zu-Han Gu, James R. Leger, and Sing H. Lee. Optical implementation of the least-squares linear mapping technique for image classification. *J. Opt. Soc. Am.*, 72(6):787–793, Jun 1982.

[95] Esperanza Guerra-Rosas and Josue Alvarez Borrego. Methodology for diagnosing of skin cancer on images of dermatologic spots by spectral analysis. *Biomed. Opt. Express*, 6(10):3876–3891, Oct 2015.

[96] Laurent Guibert, Gilles Keryer, Alain Servel, Mondher Attia, Harry S. MacKenzie, Pierre Pellat-Finet, and Jean-Louis M. de Bougrenet de la Tocnaye. On-board optical joint transform correlator for real-time road sign recognition. *Optical Engineering*, 34(1):135–143, 1995.

[97] H. S. Hinton. Photonic switching in communications systems. *J. Phys. Colloques*, 49:C2–5–C2–10, 1988.

[98] Peter Hansson and Goran Manneberg. Fourier optic characterization of paper surfaces. *Optical Engineering*, 36:35–39, 1997.

[99] Tsutomu Hara, Yoshiharu Ooi, Yoshiji Suzuki, and Ming H. Wu. Transfer characteristics of the microchannel spatial light modulator. *Appl. Opt.*, 28(22):4781–4786, Nov 1989.

[100] R. Herold and K. Leib. Image rotation in optical correlators through rotational devices. *Grumman Aerospace Corp Report*, Jan 1977.

[101] Charles F. Hester and David Casasent. Multivariant technique for multiclass pattern recognition. *Appl. Opt.*, 19(11):1758–1761, Jun 1980.

[102] Rolph Hey, Bertrand Noharet, and Henrik Sjoberg. Internet remote-controlled optical correlator based on 256 x 256 FLC spatial light modulators. *Journal of Optics A: Pure and Applied Optics*, 1(2):307–309, 1999.

[103] John A. Hoffnagle and C. Michael Jefferson. Design and performance of a refractive optical system that converts a Gaussian to a flattop beam. *Appl. Opt.*, 39(30):5488–5499, Oct 2000.

[104] Larry J. Hornbeck. Deformable-mirror spatial light modulators. *Proc. SPIE*, 1150:86–103, 1990.

[105] J. L. Horner. Clarification of Horner efficiency. *Appl. Opt.*, 31:4629, August 1992.

[106] J. L. Horner and C. K. Makekau. Compact 2f optical correlator, December 17 1991. US Patent 5,073,006.

[107] J. L. Horner and R. A. Soref. Phase-dominant spatial light modulators. *Electronics Letters*, 24(10):626–627, May 1988.

[108] Joseph L. Horner and Harmut O. Bartelt. Two-bit correlation. *Appl. Opt.*, 24(18):2889–2893, Sep 1985.

[109] Joseph L. Horner and Peter D. Gianino. Phase-only matched filtering. *Appl. Opt.*, 23(6):812–816, Mar 1984.

[110] Joseph L. Horner and Peter D. Gianino. Applying the phase-only filter concept to the synthetic discriminant function correlation filter. *Appl. Opt.*, 24(6):851–855, Mar 1985.

[111] Joseph L. Horner and Peter D. Gianino. Signal-dependent phase distortion in optical correlators. *Appl. Opt.*, 26(12):2484–2490, Jun 1987.

[112] T.D. Hudson, J.C. Kirsch, and D.A. Gregory. Comparison of optically-addressed spatial light modulators. Technical report, U.S. Army Missile Command, May 1992.

[113] K. Ito, A. Morita, T. Aoki, T. Higuchi, H. Nakajima, and K. Kobayashi. A fingerprint recognition algorithm using phase-based image matching for low-quality fingerprints. In *IEEE International Conference on Image Processing 2005*, volume 2, pages II–33–6, Sept 2005.

[114] K. Ito, H. Nakajima, K. Kobayashi, T. Aoki, and T. Higuchi. A fingerprint matching algorithm using phase-only correlation. *IEICE Transactions*, E87-A no. 3:682–691, March 2004.

[115] Jürgen Jahns and Alan Huang. Planar integration of free-space optical components. *Appl. Opt.*, 28(9):1602–1605, May 1989.

[116] K. Janschek, T. Boge, S. Dyblenko, and V. Tchernykh. Image Based Attitude Determination Using an Optical Correlator. In B. Schürmann, editor, *Spacecraft Guidance, Navigation and Control Systems*, volume 425 of *ESA Special Publication*, pages 487–492. February 2000.

[117] Bahram Javidi. Nonlinear joint power spectrum based optical correlation. *Appl. Opt.*, 28(12):2358–2367, Jun 1989.

[118] Bahram Javidi. Optical processing and pattern recognition applied to security and anticounterfeiting. *Proc. SPIE*, 10317, 2017.

[119] Bahram Javidi and Francis Chan. *Image Recognition and Classification*, chapter 6, pages 189–216. CRC Press, 2002.

[120] Bahram Javidi and Chung-Jung Kuo. Joint transform image correlation using a binary spatial light modulator at the Fourier plane. *Appl. Opt.*, 27(4):663–665, Feb 1988.

[121] Bahram Javidi, Osamu Matoba, and Enrique Tajahuerce. *Image Recognition and Classification*, chapter 8, pages 245–277. CRC Press, 2002.

[122] Bahram Javidi and Jianping Wang. Thresholding effects on the performance of the binary nonlinear joint transform correlators. *Proc. SPIE*, 1347:385–393, 1990.

[123] D. Jenkins and R. Winston. Tailored reflectors for illumination. *Appl. Opt.*, 35(10):1669–1672, Apr 1996.

[124] J.L. Jewell and A. Scherer. Surface emitting semiconductor laser, 1990. US Patent 4,949,350.

[125] D. Joyeux and S. Lowenthal. Optical Fourier transform: what is the optimal setup? *Appl. Opt.*, 21(23):4368–4372, Dec 1982.

[126] Richard D. Juday. Optimal realizable filters and the minimum Euclidean distance principle. *Appl. Opt.*, 32(26):5100–5111, Sep 1993.

[127] R. R. Kallman and D. H. Goldstein. Invariant phase-only filters for phase-encoded inputs. *Proc. SPIE*, 1564:330–347, November 1991.

[128] R. R. Kallman and D. H. Goldstein. Phase-encoding input images for optical pattern recognition. *Optical Engineering*, 33(6):1806–1812, 1994.

[129] Robert R. Kallman. The design of Phase-only Filters for Optical correlators. Technical Report AFATL-TR-90-63, Department of Mathematics University of North Texas, July 1990.

[130] H. Kawamoto. The history of liquid-crystal displays. *Proceedings of the IEEE*, 90(4):460–500, Apr 2002.

[131] R. Kerekes and B. V. K. V. Kumar. Enhanced video-based target detection using multi-frame correlation filtering. *IEEE Transactions on Aerospace and Electronic Systems*, 45(1):289–307, Jan 2009.

[132] M. Khorasaninejad, W. T. Chen, R. C. Devlin, J. Oh, A. Y. Zhu, and F. Capasso. Planar Lenses at Visible Wavelengths. *ArXiv e-prints*, May 2016.

[133] A. Kiessling. Study of real-time converter of an incoherent image into a transparency: Optical to optical converter/technology study. Technical report, April 1978.

[134] H. Kim, B. Yang, and B. Lee. Iterative Fourier transform algorithm with regularization for the optimal design of diffractive optical elements. *J Opt Soc Am A*, 21(12):2353–65, 2004.

[135] Seung-Woo Kim, SangYoon Lee, and Dong-Seon Yoon. Rapid pattern inspection of shadow masks by machine vision integrated with Fourier optics. *Optical Engineering*, 36, 1997.

[136] S. Kirkpatrick, C. D. Gelatt, and M. P. Vecchi. Optimization by simulated annealing. *Science*, 220(4598):671–680, 1983.

[137] James C. Kirsch, Harold R. Bagley, and Jeffrey A. Sloan. Dual-channel solid block optical correlator. *Proc. SPIE*, 2026:194–202, 1993.

[138] M. G. Kirzhner, M. Klebanov, V. Lyubin, N. Collings, and I. Abdulhalim. Liquid crystal high-resolution optically addressed spatial light modulator using a nanodimensional chalcogenide photosensor. *Opt Lett*, 39(7):2048–2051, 2014.

[139] Sherif Kishk and Bahram Javidi. Improved resolution 3D object sensing and recognition using time multiplexed computational integral imaging. *Opt. Express*, 11(26):3528–3541, Dec 2003.

[140] Yuji Kobayashi, Haruyoshi Toyoda, Naohisa Mukohzaka, Narihiro Yoshida, and Tsutomu Hara. Fingerprint identification by an optical joint transform correlation system. *Optical Review*, 3(6):A403, Nov 1996.

[141] I. N. Kompanets, A. V. Parfenov, and Yurii M. Popov. Spatial modulation of light in a photosensitive structure composed of a liquid crystal and an insulated gallium arsenide crystal. *Soviet Journal of Quantum Electronics*, 9(8):1070, 1979.

[142] A. Korpel, H. H. Lin, and D. J. Mehrl. Convenient operator formalism for Fourier optics and inhomogeneous and nonlinear wave propagation. *J. Opt. Soc. Am. A*, 6(5):630–635, May 1989.

[143] V.V. Kotlyar, P.G. Seraphimovich, and V.A. Soifer. An iterative algorithm for designing diffractive optical elements with regularization. *Optics and Lasers in Engineering*, 29(45):261–268, 1998.

[144] B. Kress and P. Meyrueis. *Digital diffractive optics : an introduction to planar diffractive optics and related technology.* John Wiley & Sons, Ltd, 2000.

[145] B. V. K. Vijaya Kumar. Minimum-variance synthetic discriminant functions. *J. Opt. Soc. Am. A*, 3(10):1579–1584, Oct 1986.

[146] B. V. K. Vijaya Kumar and Zouhir Bahri. Phase-only filters with improved signal to noise ratio. *Appl. Opt.*, 28(2):250–257, Jan 1989.

[147] B. V. K. Vijaya Kumar and D. Casasent. Binarization effects in a correlator with noisy input data. *Appl. Opt.*, 20(8):1433–1437, Apr 1981.

[148] B. V. K. Vijaya Kumar, Wei Shi, and Charles Hendrix. Phase-only filters with maximally sharp correlation peaks. *Opt. Lett.*, 15(14):807–809, Jul 1990.

[149] Fabrizio Lamberti, Rocco Santomo, Andrea Sanna, and Paolo Montuschi. Intensity variation function and template matching-based pedestrian tracking in infrared imagery with occlusion detection and recovery. *Optical Engineering*, 54, 2015.

[150] M. Lapisa, F. Zimmer, F. Niklaus, A. Gehner, and G. Stemme. CMOS-Integrable Piston-Type Micro-Mirror Array for Adaptive Optics Made of Mono-Crystalline Silicon using 3-D Integration. In *2009 IEEE 22nd International Conference on Micro Electro Mechanical Systems*, pages 1007–1010, Jan 2009.

[151] Alexander Laskin, Vadim Laskin, and Aleksei Ostrun. Square shaped flat-top beam in refractive beam shapers. *Proc. SPIE*, 9581, 2015.

[152] Yann LeCun, Yoshua Bengio, and Geoffrey Hinton. Deep learning. *Nature*, 521(7553):436–444, May 2015.

[153] John N. Lee and Arthur D. Fisher. Device developments for optical information processing. volume 69 of *Advances in Electronics and Electron Physics*, pages 115–173. Academic Press, 1987.

[154] James R. Leger and Sing H. Lee. Image classification by an optical implementation of the Fukunaga–Koontz transform. *J. Opt. Soc. Am.*, 72(5):556–564, May 1982.

[155] Emmett N. Leith and Juris Upatnieks. Wavefront reconstruction with diffused illumination and three-dimensional objects. *J. Opt. Soc. Am.*, 54(11):1295–1301, Nov 1964.

[156] L. B. Lesem, P. M. Hirsch, and J. A. Jordan. The kinoform: A new wavefront reconstruction device. *IBM J. Res. Dev.*, 13(2):150–155, March 1969.

[157] M. J. Lighthill. *An Introduction to Fourier Analysis and Generalised Functions.* Cambridge Monographs on Mechanics. Cambridge University Press, 1958.

[158] S. Lindell, G. Shapiro, K. Weil, Flannery, J. Levy, and Wei Qian. Development of mammogram computer-aided diagnosis systems using optical processing technology. In *Proceedings 29th Applied Imagery Pattern Recognition Workshop*, pages 173–179, 2000.

[159] Scott D. Lindell and William B. Hahn, Jr. Overview of the Martin Marietta transfer of optical processing to systems (TOPS) optical correlation program. *Proc. SPIE*, 1701:21–30, 1992.

[160] E. H. Linfoot. *Fourier Methods in Optical Image Evaluation*. The Focal Press, 1964.

[161] Tomasz Lis, Jessica Gaskin, John Jasper, and Don A. Gregory. Digital optical correlator x-ray telescope alignment monitoring system. *Optical Engineering*, 57, 2018.

[162] A. W. Lohmann and D. P. Paris. Binary Fraunhofer Holograms, Generated by Computer. *Appl. Opt.*, 6(10):1739–1748, Oct 1967.

[163] Adolf W. Lohmann, Rainer G. Dorsch, David Mendlovic, Carlos Ferreira, and Zeev Zalevsky. Space–bandwidth product of optical signals and systems. *J. Opt. Soc. Am. A*, 13(3):470–473, Mar 1996.

[164] David G. Lowe. Object Recognition from Local Scale-Invariant Features. *Proc. of the International Conference on Computer Vision ICCV, Corfu*, pages 1150–1157, 1999.

[165] Minhua Lu. Nematic liquid-crystal technology for Si wafer-based reflective spatial light modulators. *Journal of the Society for Information Display*, 10(1):37–47, 2002.

[166] J.R. Lucas, A.M. Pollack, and S.A. Mills. Reflective optical correlator with a folded asymmetrical optical axis, May 10 1994. US Patent 5,311,359.

[167] A. V. Lugt. Operational notation for the analysis and synthesis of optical data-processing systems. *Proceedings of the IEEE*, 54(8):1055–1063, Aug 1966.

[168] A. Mahalanobis and Bhagavatula Vijaya Kumar. Optimality of the maximum average correlation height filter for detection of targets in noise. *Optical Engineering*, 36:2642–2648, 1997.

[169] Abhijit Mahalanobis, B. V. K. Vijaya Kumar, and David Casasent. Minimum average correlation energy filters. *Appl. Opt.*, 26(17):3633–3640, Sep 1987.

[170] Abhijit Mahalanobis, B. V. K. Vijaya Kumar, Sewoong Song, S. R. F. Sims, and J. F. Epperson. Unconstrained correlation filters. *Appl. Opt.*, 33(17):3751–3759, Jun 1994.

[171] Abhijit Mahalanobis, Robert Stanfill, and Kenny Chen. A Bayesian approach to activity detection in video using multi-frame correlation filters. *Proc. SPIE*, 8049, 2011.

[172] Davide Maltoni, Dario Maio, Anil K. Jain, and Salil Prabhakar. *Handbook of Fingerprint Recognition*. Springer Publishing Company, Incorporated, 2nd edition, 2009.

[173] Tariq Manzur, John Zeller, and Steve Serati. Optical correlator based target detection, recognition, classification, and tracking. *Appl. Opt.*, 51(21):4976–4983, Jul 2012.

[174] Yoshinori Matsui, Munenori Takumi, Haruyoshi Toyoda, and Seiichiro Mizuno. A 3.2 khz, stereo sensing module using two profile sensors. In *MVA, IAPR Conference on Machine Vision Applications*, 2009.

[175] C. Maurer, A. Jesacher, S. Bernet, and M. Ritsch-Marte. What spatial light modulators can do for optical microscopy. *Laser and Photonics Reviews*, 5(1):81–101, 2011.

[176] C. Maurer, A. Jesacher, S. Furhapter, S. Bernet, and M. Ritsch-Marte. Upgrading a microscope with a spiral phase plate. *Journal of Microscopy*, 230(1):134–142, 2008.

[177] S. Maze and P. Refregier. Optical correlation: influence of the coding of the input image. *Appl. Opt.*, 33:6788–6796, October 1994.

[178] Gregor J. McDonald, Meirion F. Lewis, and Rebecca Wilson. A high-speed readout scheme for fast optical correlation-based pattern recognition. *Proc. SPIE*, 5616:85–92, 2004.

[179] Stephen D. Mellin and Gregory P. Nordin. Limits of scalar diffraction theory and an iterative angular spectrum algorithm

for finite aperture diffractive optical element design. *Opt. Express*, 8(13):705–722, Jun 2001.

[180] J. L. Metz and K. M. Johnson. Optically Computing the Hit Miss Transform for an Automated Cervical Smear Screening System. *Appl. Opt.*, 39:803–813, February 2000.

[181] S. Mias. *Fabrication and characterisation of optically addressed spatial light modulators*. PhD Thesis, University of Cambridge, May 2004.

[182] M. S. Millan, E. Perez, and K. Chalasinska-Macukow. Pattern recognition with variable discrimination capability by dual nonlinear optical correlation. *Optics Communications*, 161:115–122, March 1999.

[183] Gladys Minguez-Vega, Matthias Gruber, Jürgen Jahns, and Jesus Lancis. Achromatic optical Fourier transformer with planar-integrated free-space optics. *Appl. Opt.*, 44(2):229–235, Jan 2005.

[184] K. Miyazawa, K. Ito, T. Aoki, K. Kobayashi, and H. Nakajima. An effective approach for iris recognition using phase-based image matching. *IEEE Transactions on Pattern Analysis and Machine Intelligence*, 30(10):1741–1756, Oct 2008.

[185] S.R. Morrison. A new type of photosensitive junction device. *Solid-State Electronics*, 6(5):485–494, 1963.

[186] Thibault Napoleon and Ayman Alfalou. Local binary patterns preprocessing for face identification/verification using the VanderLugt correlator. *Proc. SPIE*, 9094, 2014.

[187] M. Nazarathy and J. Shamir. Fourier optics described by operator algebra. *J. Opt. Soc. Am.*, 70(2):150–159, Feb 1980.

[188] J. B. Nelson. Multivariant technique for multiclass pattern recognition: comment. *Appl. Opt.*, 20(1):8–9, Jan 1981.

[189] N.J. New and R. Todd. Alignment method, July 14 2016. WO Patent App. PCT/GB2015/054,058.

[190] Randall K. Nichols. *Icsa Guide to Cryptography*. McGraw-Hill Professional, 1998.

[191] Fredrik Nikolajeff, Jörgen Bengtsson, Michael Larsson, Mats Ekberg, and Sverker Hard. Measuring and modeling the proximity effect in direct-write electron-beam lithography kinoforms. *Appl. Opt.*, 34(5):897–903, Feb 1995.

[192] Fredrik Nikolajeff, Stellan Jacobsson, Sverker Hard, Ake Billman, Lars Lundbladh, and Curt Lindell. Replication of continuous-relief diffractive optical elements by conventional compact disc injection-molding techniques. *Appl. Opt.*, 36(20):4655–4659, Jul 1997.

[193] M.J. O'Callaghan. Optical correlator having multiple active components formed on a single integrated circuit, June 12 2001. US Patent 6,247,037.

[194] Donald C. O'Shea, Thomas J. Suleski, Alan D. Kathman, and Dennis W. Prather. *Diffractive optics: design, fabrication, and test.* SPIE Press Washington, DC, USA, 2004.

[195] Rafael Gil Otero, Craig J. Moir, Theodore Lim, Gordon A. Russell, and John Fraser Snowdon. Free-space optical interconnected topologies for parallel computer application and experimental implementation using rapid prototyping techniques. *Optical Engineering*, 45, 2006.

[196] Yousri Ouerhani, Ayman Alfalou, and Christian Brosseau. Road mark recognition using HOG-SVM and correlation. *Proc. SPIE*, 10395, 2017.

[197] Jae-Hyeung Park, Joohwan Kim, and Byoungho Lee. Three-dimensional optical correlator using a sub-image array. *Opt. Express*, 13(13):5116–5126, Jun 2005.

[198] William A. Parkyn and David G. Pelka. New TIR lens applications for light-emitting diodes. *Proc. SPIE*, 3139, 1997.

[199] Robert M. Pasternack, Bryan Rabin, Jing-Yi Zheng, and Nada N. Boustany. Quantifying subcellular dynamics in apoptotic cells with two-dimensional Gabor filters. *Biomed. Opt. Express*, 1(2):720–728, Sep 2010.

[200] Robert M. Pasternack, Jing-Yi Zheng, and Nada N. Boustany. Detection of mitochondrial fission with orientation-dependent optical Fourier filters. *Cytometry Part A*, 79A(2):137–148, 2011.

[201] R. Patnaik and D. Casasent. Kernel synthetic discriminant function (SDF) filters for fast object recognition. *Proc. SPIE*, 7340, 2009.

[202] A.R. Peaker. Light-emitting diodes. *IEE Proceedings A (Physical Science, Measurement and Instrumentation, Management and Education, Reviews)*, 127:202–210(8), April 1980.

[203] Sergio Pinto-Fernandez and Victor H. Diaz-Ramirez. Improvement of facial recognition with composite correlation filters designed with combinatorial optimization. *Proc. SPIE*, 8498, 2012.

[204] N. I. Pletneva, I. E. Morichev, F. L. Vladimirov, and L. I. Basyaeva. Spatial and temporal light modulator based on a photoconductor liquid crystal structure with a glass fiber component. *Soviet Journal of Quantum Electronics*, 13(9):1253, 1983.

[205] A. G. Poleshchuk, V. P. Korolkov, and R. K. Nasyrov. Diffractive optical elements: fabrication and application. *Proc. SPIE*, 9283, 2014.

[206] Mary C. Potter, Brad Wyble, Carl Erick Hagmann, and Emily S. McCourt. Detecting meaning in RSVP at 13 ms per picture. *Attention, Perception, and Psychophysics*, 76(2):270–279, 2014.

[207] Demetri Psaltis, Eung G. Paek, and Santosh S. Venkatesh. Optical image correlation with a binary spatial light modulator. *Optical Engineering*, 23(6), 1984.

[208] Jin Qin, Siqi Tang, Congying Han, and Tiande Guo. Partial fingerprint matching via phase-only correlation and deep convolutional neural network. In *Neural Information Processing: 24th International Conference, ICONIP 2017, Guangzhou, China, November 14–18, 2017, Proceedings, Part VI*, pages 602–611. Springer International Publishing, Cham, 2017.

[209] F. Quercioli, B. Tiribilli, A. Mannoni, and S. Acciai. Optomechanics with LEGO. *Appl. Opt.*, 37(16):3408–3416, Jun 1998.

[210] Henri Rajbenbach, Sambath Bann, Philippe Refregier, Pascal Joffre, Jean-Pierre Huignard, Hermann-Stephan Buchkremer, Arne Skov Jensen, Erling Rasmussen, Karl-Heinz Brenner, and Garry Lohman. Compact photorefractive correlator for robotic applications. *Appl. Opt.*, 31(26):5666–5674, Sep 1992.

[211] Miquel Rallo, Marja S. Millan, and Jaume Escofet. Referenceless segmentation of flaws in woven fabrics. *Appl. Opt.*, 46(27):6688–6699, Sep 2007.

[212] J A Ratcliffe. Some aspects of diffraction theory and their application to the ionosphere. *Reports on Progress in Physics*, 19(1):188, 1956.

[213] G. Ravichandran and D. P. Casasent. Noise and discrimination performance of the MINACE optical correlation filter. *Proc. SPIE*, 1471:233–248, 1991.

[214] Saad Rehman, Farhan Riaz, Ali Hassan, Muwahida Liaquat, and Rupert Young. Human detection in sensitive security areas through recognition of omega shapes using MACH filters. *Proc. SPIE*, 9477, 2015.

[215] S. Reinhorn, Y. Amitai, and A. A. Friesem. Compact planar optical correlator. *Opt. Lett.*, 22(12):925–927, Jun 1997.

[216] Harald R. Ries and Roland Winston. Tailored edge-ray reflectors for illumination. *J. Opt. Soc. Am. A*, 11(4):1260–1264, Apr 1994.

[217] Alan L. Rockwood, David K. Crockett, James R. Oliphant, and Kojo S. J. Elenitoba-Johnson. Sequence alignment by cross-correlation. *Journal of Biomolecular Techniques: JBT*, 16(4):453–458, December 2005.

[218] J. F. Rodolfo, H. J. Rajbenbach, and J.-P. Huignard. Performance of a photorefractive joint transform correlator for fingerprint identification. *Optical Engineering*, 34, April 1995.

[219] Joseph Rosen. Three-dimensional electro-optical correlation. *J. Opt. Soc. Am. A*, 15(2):430–436, Feb 1998.

[220] Frank Rosenblatt. *Principles of Neurodynamics: Perceptrons and the Theory of Brain Mechanisms*. Spartan Books, 1962.

[221] Azriel Rosenfeld. *Picture Processing by Computer*. Academic, New York, NY, USA, 1969.

[222] Filippus Stefanus Roux. Diffractive optical implementation of rotation transform performed by using phase singularities. *Appl. Opt.*, 32(20):3715–3719, Jul 1993.

[223] Filippus Stefanus Roux. Implementation of general point transforms with diffractive optics. *Appl. Opt.*, 32(26):4972–4978, Sep 1993.

[224] Filippus Stefanus Roux. Diffractive optical Hough transform implemented with phase singularities. *Appl. Opt.*, 33(14):2955–2959, May 1994.

[225] Filippus Stefanus Roux. Extracting one-dimensional wavelet features with a diffractive optical inner-product transform. *Appl. Opt.*, 35(23):4610–4614, Aug 1996.

[226] Filippus Stefanus Roux. Rotation- and scale-invariant feature extraction by a diffractive optical inner-product transform. *Appl. Opt.*, 35(11):1894–1899, Apr 1996.

[227] Per Rudquist. *Handbook of Visual Display Technology*, chapter Smectic LCD Modes, pages 1445–1467. Springer Berlin Heidelberg, 2012.

[228] J.A. Sachs, H.V. Goetz, D. Keith, L. Li, S.H. Linn, A. Parfenov, S.E. Brice, T.J. Scheffer, J.A. Van Vechten, and J. Xue. Transmissive, optically addressed, photosensitive spatial light modulators and color display systems incorporating same, August 2011. US Patent 7,990,600.

[229] M. Savvides, B. V. K. V. Kumar, and P. K. Khosla. Eigenphases vs eigenfaces. In *Proceedings of the 17th International Conference on Pattern Recognition, 2004*, volume 3, pages 810–813, Aug 2004.

[230] Marija Scholl and James W. Scholl. Optical processing for range and attitude determination. *Proc. SPIE*, 1694:23–28, 1992.

[231] Marija S. Scholl. Application of a hybrid digital-optical cross-correlator as a semi-autonomous vision system. *Proc. SPIE*, 1772:128–135, 1992.

[232] S. D. Searle, A. G. Levenston, C. Stace, H. White, and S. Parker. Application of an optical correlator to industrial inspection. *Proc. SPIE*, 0654:178–181, 1986.

[233] Steven A. Serati, Teresa K. Ewing, Roylnn A. Serati, Kristina M. Johnson, and Darren M. Simon. Programmable 128 x 128 ferroelectric liquid crystal spatial light modulator compact correlator. *Proc. SPIE*, 1959:55–68, 1993.

[234] Thomas Serre, Aude Oliva, and Tomaso Poggio. A feedforward architecture accounts for rapid categorization. *Proceedings of the National Academy of Sciences*, 104(15):6424–6429, 2007.

[235] James H. Sharp, Nick E. MacKay, Pei C. Tang, Ian A. Watson, Brian F. Scott, David M. Budgett, Chris R. Chatwin, Rupert C. D. Young, Sylvie Tonda, Jean-Pierre Huignard, Tim G. Slack, Neil Collings, Ali-Reza Pourzand, Marcus Duelli, Aldo Grattarola, and Carlo Braccini. Experimental systems implementation of a hybrid optical–digital correlator. *Appl. Opt.*, 38(29):6116–6128, Oct 1999.

[236] Pawan Kumar Shrestha, Young Tea Chun, and Daping Chu. A high-resolution optically addressed spatial light modulator based on ZnO nanoparticles. *Light Sci Appl*, 4:e259, 2015.

[237] Di Si and Jin Jian. A compact finger-vein identification system based on infrared imaging. *Proc. SPIE*, 10462, 2017.

[238] Terence Sim, Simon Baker, and Maan Bsat. The CMU Pose, Illumination, and Expression (PIE) Database of Human Faces. Technical Report CMU-RI-TR-01-02, Robotics Institute, Carnegie Mellon University, Pittsburgh, PA, January 2001.

[239] D.M. Simon and S.A. Serati. Optical correlator using ferroelectric liquid crystal spatial light modulators and Fourier transform lenses, May 23 1995. US Patent 5,418,380.

[240] Henrik J. Sjoberg, Bertrand Noharet, Lech Wosinski, and Rolph Hey. Compact optical correlator: pre-processing and filter encoding strategies applied to images with varying illumination. *Optical Engineering*, 37(4):1316–1324, 1998.

[241] Jeffrey A. Sloan and Donald W. Small. Design and fabrication of a miniaturized optical correlator. *Optical Engineering*, 32:3307–3315, 1993.

[242] H. M. Smith, editor. *Holographic Recording Materials (Topics in applied physics)*. Springer-Verlag, 1977.

[243] W.J. Smith. *Modern Optical Engineering, 4th Ed.* McGraw-Hill Education, 2007.

[244] L.M. Soroko. *Holography and Coherent Optics.* Springer US, 1980.

[245] K. T. Stalker, P. A. Molley, and F. M. Dickey. Real-time optical processor for synthetic aperture radar image formation. *Proc. SPIE*, 0789:96–104, 1987.

[246] T. C. Strand. Optical three-dimensional sensing for machine vision. *Optical Engineering*, 24(1):33–40, 1985.

[247] Marija Strojnik, Michael S. Shumate, Richard L. Hartman, Jeffrey A. Sloan, and Donald W. Small. Miniaturized optical correlator. *Proc. SPIE*, 1347:186–198, 1990.

[248] S. I. Sudharsanan, A. Mahalanobis, and M. K. Sundareshan. Unified framework for the synthesis of synthetic discriminant functions with reduced noise variance and sharp correlation structure. *Optical Engineering*, 29, 1990.

[249] Y. Sugimoto, S. Ozawa, and N. Inaba. Spaceborne synthetic aperture radar signal processing using FPGAs. *Proc. SPIE*, 10430, 2017.

[250] Jing-Gao Sui, Wu-Sheng Tang, Xiao-Ya Zhang, Chen-Cheng Feng, and Hui Jia. A novel high-speed opto-electronic hybrid correlator for recognition and tracking. *Proc. SPIE*, 8559, 2012.

[251] Sarun Sumriddetchkajorn and Kosom Chaitavon. A Fourier-optics-based non-invasive and vibration-insensitive micron-size object analyzer for quality control assessment. *Proc. SPIE*, 6995, 2008.

[252] K. Takita, T. Aoki, Y. Sasaki, T. Higuchi, and K. Kobayashi. High-accuracy subpixel image registration based on phase-only correlation. *IEICE Transactions*, E86-A no. 8:1925–1934, August 2003.

[253] Maxine Tan, Wei Qian, Jiantao Pu, Hong Liu, and Bin Zheng. A new approach to develop computer-aided detection schemes of digital mammograms. *Physics in Medicine and Biology*, 60(11):4413, 2015.

[254] V. Tchernykh, M. Beck, and K. Janschek. An embedded optical flow processor for visual navigation using optical correlator technology. In *2006 IEEE/RSJ International Conference on Intelligent Robots and Systems*, pages 67–72, Oct 2006.

[255] V. Tchernykh, M. Beck, and K. Janschek. Optical flow navigation for an outdoor UAV using a wide angle mono camera and DEM matching. *IFAC Proceedings Volumes*, 39(16):590–595, 2006. 4th IFAC Symposium on Mechatronic Systems.

[256] Duncan J. Telfer and Zhongqiang Li. Fast texture segmentation by Fourier optic implementation of Gaussian smoothed fan filters. *Proc. SPIE*, 1704, 1992.

[257] Jiangmin Tian, Guoyou Wang, Jianguo Liu, and Yuanchun Xia. Chinese license plate character segmentation using multiscale template matching. *Journal of Electronic Imaging*, 25, 2016.

[258] Q. Tian, Y. Fainman, and Sing H. Lee. Comparison of statistical pattern-recognition algorithms for hybrid processing. II. Eigenvector-based algorithm. *J. Opt. Soc. Am. A*, 5(10):1670–1682, Oct 1988.

[259] P. Topiwala and A. Nehemiah. Real-time multi-sensor based vehicle detection using MINACE filters. *Proc. SPIE*, 6574, 2007.

[260] G. Tricoles. Computer generated holograms: an historical review. *Appl. Opt.*, 26(20):4351–4360, Oct 1987.

[261] Matthew Turk and Alex Pentland. Eigenfaces for recognition. *J. Cognitive Neuroscience*, 3(1):71–86, January 1991.

[262] Juris Upatnieks and James O. Abshier. Compact coherent optical correlator system. Technical Report AD-A200 769, Environmental Research Institute of Michigan (ERIM), May 1988.

[263] P.H. van Cittert. Degree of coherence. *Physica*, 24(5):505 – 507, 1958.

[264] A.B. vander Lugt. Signal detection by complex spatial filtering. Technical Report 2900-394-T, University of Michigan, Institute of Science and Technology, July 1963.

[265] V. N. Vapnik and A. Ya. Chervonenkis. On the uniform convergence of relative frequencies of events to their probabilities. In Vladimir Vovk, Harris Papadopoulos, and Alexander Gammerman, editors, *Measures of Complexity: Festschrift for Alexey Chervonenkis*, pages 11–30. Springer International Publishing, Cham, 2015.

[266] W. B. Veldkamp. Laser beam profile shaping with interlaced binary diffraction gratings. *Appl. Opt.*, 21(17):3209–3212, Sep 1982.

[267] Wilfrid B. Veldkamp and Carol J. Kastner. Beam profile shaping for laser radars that use detector arrays. *Appl. Opt.*, 21(2):345–356, Jan 1982.

[268] Bhagavatula Vijaya Kumar and Charles D. Hendrix. Choice of threshold line angle for binary phase-only filters. *Proc. SPIE*, 1959:170–185, 1993.

[269] Paul Viola and Michael Jones. Robust real-time object detection. *International Journal of Computer Vision*, 57:137–154, 2004.

[270] K. von Bieren. Lens design for Optical Fourier Transform systems. *Appl. Opt.*, 10(12):2739–2742, Dec 1971.

[271] Daniel Vukobratovich. Modular optical alignment. *Proc. SPIE*, 3786:427–438, 1999.

[272] Li Wang and Garret Moddel. Resolution limits from charge transport in optically addressed spatial light modulators. *Journal of Applied Physics*, 78(12):6923–6935, 1995.

[273] Qin Wang, Andy Z. Zhang, Susanne Almqvist, Stephane Junique, Bertrand Noharet, Duncan Platt, Michael Salter, and Jan Y. Andersson. Recent developments in electroabsorption modulators at Acreo Swedish ICT. *Proc. SPIE*, 9362, 2015.

[274] Shunli Wang, Liangcai Cao, Huarong Gu, Qingsheng He, Claire Gu, and Guofan Jin. Channel analysis of the volume holographic correlator for scene matching. *Opt. Express*, 19(5):3870–3880, Feb 2011.

[275] Shunli Wang, Qiaofeng Tan, Liangcai Cao, Qingsheng He, and Guofan Jin. Multi-sample parallel estimation in volume holographic correlator for remote sensing image recognition. *Opt. Express*, 17(24):21738–21747, Nov 2009.

[276] Edward R. Washwell, Rolin J. Gebelein, Gregory O. Gheen, David Armitage, and Mark A. Handschy. Miniature hybrid optical correlators: device and system issues. *Proc. SPIE*, 1297:64–71, 1990.

[277] Eriko Watanabe and Kashiko Kodate. Fast face-recognition optical parallel correlator using high accuracy correlation filter. *Optical Review*, 12(6):460–466, 2005.

[278] Eriko Watanabe, Anna Naito, and Kashiko Kodate. Ultra-high-speed compact optical correlation system using holographic disc. *Proc. SPIE*, 7442, 2009.

[279] P. Watson, P. J. Bos, J. Gandhi, Y. Ji, and M. Stefanov. Optimization of bend cells for field-sequential color microdisplay applications. *SID Symposium Digest of Technical Papers*, 30(1):994–997, 1999.

[280] C. S. Weaver and J. W. Goodman. A technique for optically convolving two functions. *Appl. Opt.*, 5(7):1248–1249, Jul 1966.

[281] H. Weyl. Ausbreitung elektromagnetischer wellen ber einem ebenen leiter. *Annalen der Physik*, 365(21):481–500, 1919.

[282] Norbert Wiener. *Extrapolation, Interpolation, and Smoothing of Stationary Time Series*. The MIT Press, 1964.

[283] Roland Winston and Weiya Zhang. Novel aplanatic designs. *Opt. Lett.*, 34(19):3018–3019, Oct 2009.

[284] C.G. Wynne. Simple Fourier transform lenses II. *Optics Communications*, 12(3):270–274, 1974.

[285] Jingqun Xi. High efficiency directional light source using lens optics, June 2 2015. US Patent 9,046,241.

[286] Rupert C. D. Young, Christopher R. Chatwin, and Brian F. Scott. High-speed hybrid optical/digital correlator system. *Optical Engineering*, 32(10):2608–2615, 1993.

[287] Zeev Zalevsky, Rainer G. Dorsch, and David Mendlovic. Gerchberg–Saxton algorithm applied in the fractional Fourier or the Fresnel domain. *Opt. Lett.*, 21(12):842–844, Jun 1996.

[288] Xu Zeng, Jian Bai, Changlun Hou, and Guoguang Yang. Compact optical correlator based on one phase-only spatial light modulator. *Opt. Lett.*, 36(8):1383–1385, Apr 2011.

[289] Xu Zeng, Takashi Inoue, Norihiro Fukuchi, and Jian Bai. Parallel lensless optical correlator based on two phase-only spatial light modulators. *Opt. Express*, 19(13):12594–12604, Jun 2011.

[290] F. Zernike. How I discovered phase contrast. *Science*, 121(3141):345–349, 1955.

[291] Jie Zhang, Binjie Xin, and Xiangji Wu. Density measurement of yarn dyed woven fabrics based on dual-side scanning and the FFT technique. *Measurement Science and Technology*, 25(11):115007, 2014.

[292] Nikolay Zheludev. The life and times of the LED, a 100-year history. *Nature Photonics*, 1(4):189–192, 2007.

Index

Milton Keynes UK
Ingram Content Group UK Ltd.
UKHW040055071024
449327UK00019B/579